HEAT
TRANSFER
POCKET
HANDBOOK

Gulf Publishing Company
Book Division
Houston, London, Paris, Tokyo

HEAT TRANSFER POCKET HANDBOOK

Nicholas P. Cheremisinoff

Heat Transfer Pocket Handbook

Library of Congress Cataloging in Publication Data

Cheremisinoff, Nicholas P.
    Heat transfer pocket handbook.
    Includes bibliographical references and index.
 1. Title.
QC320.4.C47   1984   621.402'2   84-654
ISBN 0-87201-379-0

CONTENTS

3

CONVECTIVE HEAT TRANSFER 78

4

RADIATION HEAT TRANSFER 110

5

HEAT EXCHANGER
CLASSIFICATION 119

6

HEAT EXCHANGER DESIGN NOTES .. 137

PREFACE

The *Heat Transfer Pocket Handbook* is a short, concise guide to solving heat transfer problems. This volume, along with its companion, the *Fluid Flow Pocket Handbook,* was prepared as the engineer's helper. As a compilation of design notes, equipment scale-up criteria, useful formulas, and data, it will enable at least first-pass design analyses to be made of typical industrial problems concerned with heat transmission.

The book is not intended as a reference text but as a supplemental guide to detailed calculations and design practices. Little theory is presented, although key references are included for the user who requires in-depth coverage of topics. Each section is organized into a concise set of notes, and detailed sample calculations are included to illustrate proper use of design methods. Process engineers, in particular chemical and mechanical, will find this a helpful addition to their design notes. Students will also find this to be a valuable study guide and supplement to coursework and laboratory experiments on heat transfer.

Nicholas P. Cheremisinoff

NOTATION

A	area
a	parameter
B	parameter in Equation 8–19; boiling parameter in Chapter 10.
Bi	dimensionless Biot number
C_p, C_v	specific heats at constant pressure and volume, respectively
C_m	mean concentration
C_G, C_H	heat capacity rate
C_h, C_C	heat capacity rates
C_{sf}	nucleate boiling constant
c	speed of light
C^*	dimensionless rate term (hot to cold fluid ratio)
D	diffusivity
D_h	equivalent diameter in Chapter 1
d	diameter or distance
$E_{b,\lambda}$	emissive power
E	internal energy
Eu	Euler number
F	force in Chapter 1; geometrical factor in Chapter 4; function in Equation 10-22
Fr	Froude number
Fo	Fourier number
f	friction factor
G	mass flow per unit area
Gr	Grashof number
Gz	Graetz number
g	gravitational acceleration
g_c	conversion factor
H	enthalpy; height in Chapter 2

H_i, H_o	header parameters in Chapter 7
h	heat transfer coefficient
h_o	Planck's constant
I	resistance parameter in Chapter 2; radiation intensity in Chapter 4
i'_g	effective latent heat of vaporization
J	Richardson number
j	Colburn factor
j_D, j_F	dropwise and filmwise condensation mass flux, respectively
j_ℓ, j_g	superficial liquid and gas velocities, respectively
K	overall heat transfer coefficient in Chapter 2; velocity ratio in Chapter 7
$k(\infty)$	entrance factor in Chapter 4
k	thermal conductivity
L, l	length
LMTD	log-mean temperature difference
M	molecular weight or mass
Ma	Maragoni number
m	mass; inverse of Biot number in Chapter 2
N_m	mass transfer rate
N_{tu}	number of heat transfer units
Nu	Nusselt number
N_{tu}, N	number of transfer units
n	number of tubes in Chapter 9; dimensionless size in Chapter 2
P	pressure
Pe	Peclet number
Pr	Prandtl number
p	perimeter
Q, q	heat transfer rate
q''	heat flux
R	universal gas constant in Chapter 1; radius
R_T	thermal resistance per unit thickness
R_f	fouling factor
Ra	Rayleigh number
Ra*	modified Rayleigh number
Re	Reynolds number
r	radius; thermal resistance in Chapter 2
r_h	hydraulic radius
S	suppression factor, see Equation 10-23; cross-sectional area in Chapter 3; entropy

Sc   Schmidt number
St   Stanton number
T   absolute temperature
t   temperature
U   overall heat transfer coefficient
u_m mean velocity
V_m volume per mole
W   mass flow
x   vapor quality
x^+, x^* dimensionless distances
Z   compressibility factor
z   distance

Greek Symbols

α thermal diffusivity; absorptivity in Chapter 4
β coefficient of volumetric expansion in Chapter 1; liquid vapor fraction in Chapter 8
Γ_D flow function—see Equation 9-4; wetted perimeter in Chapter 3; pressure ratio in Chapter 8
Γ_Z condensate mass flow rate
Δ dimensionless temperature difference
δ film thickness
ϵ emissivity in Chapter 4; heat exchanger effectiveness
ϵ_{ho} temperature effectiveness
η_f fin efficiency
θ dimensionless temperature; angle
Λ coefficient defined in Equation 8-3
$\varkappa$ Boltzman constant
λ wavelength; two-phase flow parameter in Chapter 8; heat of vaporization in Chapter 3
μ fluid viscosity
ν kinematic viscosity or specific volume
ρ fluid density
σ surface tension; Stefan-Boltzmann constant in Chapter 4
τ time
τ_w wall shear stress
ϕ Lockhart-Martinelli parameter; angle in Chapter 9
χ ratio of liquid to gas single-phase pressure drops (Martinelli parameter); angle in Chapter 9
ψ two-phase flow parameter—see Equation 8-5

1

TRANSPORT PROPERTIES AND DIMENSIONLESS GROUPS

Dimensionless Groups in Fluid Flow

The following is a list of important dimensionless groups in fluid flow analysis.

Friction Factor

The Fanning friction factor is defined as the ratio of wall shear stress τ_w to the flow's kinetic energy per unit volume (i.e., the dynamic velocity head or velocity pressure):

$$f = \frac{\tau_w}{\rho u_m^2/2g_c} \tag{1-1}$$

There are several other definitions of "friction factor," and distinction should be made before its use. Another common definition is the Darcy friction factor:

$$f_D = 4f \tag{1-2}$$

Euler Number

This is defined as:

$$Eu = \frac{\Delta P}{\rho u_m^2/g_c} = \frac{\Delta P^*}{2} \tag{1-3}$$

Note that in a steady-state fully developed flow, momentum exchange at any cross-section is constant. Pressure drop then results from the

1

wall friction. For no core entrance and exit losses, Newton's second law of motion shows that:

$$\Delta P^* = \frac{\Delta P}{\rho u_m^2 / 2g_c} = f\frac{L}{r_h} \qquad (1\text{-}4)$$

where L is the length of pipe in which ΔP occurs.

Reynolds Number

This is defined as the ratio of flow momentum rate (i.e., inertia force) to viscous force.

$$Re = \frac{\rho u_m D_h}{\mu} = \frac{G D_h}{\mu} \qquad (1\text{-}5)$$

where ρ = fluid density
u_m = mean fluid velocity
μ = fluid viscosity
D_h = equivalent diameter
G = mass throughput

Froude Number

This is defined as the ratio of inertia force on a fluid element to the gravity force.

$$Fr = \frac{u_m^2}{g\ell} \qquad (1\text{-}6)$$

where ℓ is the characteristic size (height or equipment diameter).

Dimensionless Axial Distance

For the hydrodynamic entrance region, the dimensionless axial distance in the flow direction is defined as:

$$x^+ = \frac{x}{D_h Re} \qquad (1\text{-}7)$$

Note that the apparent friction factor, f_{app}, decreases with increasing x^+ values and asymptotically approaches the fully developed value f as

$x^+ \rightarrow \infty$. The apparent friction factor is used to describe developing flows. In this case, momentum rate changes at any cross-section in tube flow. The effects of wall friction and momentum rate changes in developing flows are related through f_{app}:

$$\Delta P^* = \frac{\Delta P}{\rho u_m^2/2g_c} = f_{app}\frac{L}{r_h} = f_{fd}\frac{L}{r_h} + \Delta(x) \tag{1-8}$$

where $\Delta(x)$ = incremental pressure drop number
       r_h = hydraulic radius

Rayleigh Number

This is a dimensionless group used to define criterion for the onset of Rayleigh-Bernard flow instabilities.

$$Ra = \frac{g\beta\Delta T\delta^3}{\alpha\nu} \tag{1-9}$$

where g = gravitational acceleration
       β = coefficient of volumetric expansion
       ΔT = temperature difference between wall and fluid
       δ = fluid thickness
       α = thermal diffusivity
       ν = fluid viscosity

Density gradients due to temperature or concentration variations in a stratified liquid layer produce flow instabilities when the heavier fluid particles are above the less dense particles.

Maragoni Number

Surface tension instability in two-phase flows is characterized by this group:

$$Ma = -(\partial\sigma/\partial T)_0(\Delta T)\delta/\alpha\mu \tag{1-10}$$

This instability generally occurs during mass transfer across a gas-liquid interface.

Richardson Number

This is the ratio of the buoyant force to the inertia or shear force:

$$J = \frac{g(dP/dZ)}{\rho(dU/dZ)^2} \tag{1-11}$$

When different layers of a stratified homogeneous fluid are in relative motion, instability arises because of interaction between fluid phases. The above definition ignores the viscosities of the two fluids.

Dimensionless Groups in Heat Transfer

The following is a list of important dimensionless groups in heat transfer.

Nusselt Number

This is defined as the ratio of the convective conductance h to the pure molecular thermal conductance k/D_h.

$$Nu = \frac{h}{k/D_h} = \frac{hD_h}{k} = \frac{q''D_h}{k(t_w - t_m)} \tag{1-12}$$

Note that the heat transfer coefficient h represents the convective conductance in a thermal circuit representation with the heat flux q'' as the current and $t_w - t_m$ as the potential.

The Nu strongly depends upon the thermal boundary condition and flow passage geometry in laminar flow, and is weakly dependent upon these parameters in turbulent flow. The Nusselt number is constant for thermally and hydrodynamically fully developed laminar flow. It is dependent upon $x^* = x/(D_h Pe)$, where Pe is the Peclet number, for developing laminar temperature profiles and is dependent upon x^* and Pr for developing laminar velocity and temperature profile. Nu also depends upon the Reynolds (Re) and Prandtl (Pr) numbers for fully developed turbulent flows. The Nusselt number is related to the Stanton, Prandtl, and Reynolds numbers as follows:

$$Nu = St\ Pr\ Re \tag{1-13}$$

Stanton Number

This is defined as the ratio of convected heat transfer (per unit duct surface area) to the amount virtually transferable, as if temperature equalization were attained (per unit of cross-sectional flow area):

$$St = \frac{h}{GC_p} \frac{(t_w - t_m)}{(t_w - t_m)} = \frac{h}{GC_p} \tag{1-14}$$

For large Peclet numbers (i.e., when axial fluid heat conduction is small), St is frequently preferred, instead of Nu, as a dimensionless modulus for the convective heat transfer correlation. The reason being that it can be related to the number of heat exchanger transfer units N_{tu}. Also, the behavior of St with Re is analogous to that of the Fanning friction factor f versus Re.

The Stanton number also is directly related to the number of heat transfer units of one side of the exchanger.

$$St = N_{tu} \frac{D_h}{4L} \tag{1-15}$$

The Stanton number is dependent upon the fluid Prandtl number, Reynolds number, and the flow system geometry.

Colburn Factor (j)

This is a modulus that accounts for the moderate variations in the Prandtl number. The j versus Re characteristics for a given surface is nearly independent of the flowing fluid for $0.5 \leq Pr \leq 10.0$ from laminar to turbulent flow conditions.

$$j = St\, Pr^{2/3} = \frac{Nu\, Pr^{-1/3}}{Re} \tag{1-16}$$

Prandtl Number (Pr)

This is defined as the ratio of momentum diffusivity to thermal diffusivity of the fluid.

$$Pr = \frac{\nu}{\alpha} = \frac{\mu C_p}{k} \tag{1-17}$$

The Prandtl number is solely a fluid property modulus. Typical ranges for several types of fluids are as follows: 0.001–0.03 for liquid metals, 0.2–1 for gases, 1–10 for water, 5–50 for light organic liquids, and 50–2,000 for oils.

Peclet Number (Pe)

This is defined as follows:

$$Pe = \frac{\rho C_p u_m D_h}{k} = \frac{u_m D_h}{\alpha} \tag{1-18}$$

On multiplying the numerator and denominator by the axial fluid bulk mean temperature gradient (dt_m/dx), it can be shown that

$$Pe = D_h \frac{W c_p (dt_m/dx)}{k A_0 (dt_m/dx)} \tag{1-19}$$

The Peclet number represents the relative magnitude of the thermal energy convected to the fluid (fluid enthalpy rise) to the thermal energy axially conducted within the fluid. The inverse of the Peclet number is representative of the relative importance of fluid axial heat conduction. Therefore, Pe is important for liquid metal heat exchangers, where

$$Pe = Re\ Pr \simeq \frac{\text{fluid enthalpy rise}}{\text{axial energy conduction}} \tag{1-20}$$

Dimensionless Axial Distance (x*) and Graetz Number (Gz)

The dimensionless distance in the flow direction for the thermal entrance region heat transfer is:

$$x^* = \frac{x}{D_h Pe} = \frac{x}{D_h Re Pr} \tag{1-21}$$

The Graetz number, Gz, is defined as:

$$Gz = \frac{W C_p}{kL} = \frac{PeP}{4L} = \frac{RePrP}{4L} \tag{1-22}$$

If the flow length L in Gz is considered as a length variable, then:

$$x^* = \frac{P}{4D_h} \frac{1}{Gz} \tag{1-23}$$

which reduces to $x^* = \pi/(4\ Gz)$ for the circular tube.

Grashof Number

This is a dimensionless group used to define heat transfer by natural convection:

$$Gr = \frac{gD^3}{\nu^2} \beta(t_w - t_\infty) \tag{1-24}$$

Movement of the fluid during natural convection occurs because of density differences brought about by temperature gradients. The variation of density with temperature is represented by the coefficient of thermal expansion, β. The rate of natural-convection heat transfer is also established by the magnitude of the temperature difference $t_w - t_\infty$ between the wall and fluid.

Table 1-1 summarizes the significance of the major dimensionless groups in convection heat transfer. Definitions of symbols used in the above formulas are given in Table 1-2. Figure 1-1 illustrates the correlation of the dimensionless groups for the case of air flow in a circular tube. Note the similarity in j and f versus Re characteristics. No similarity exists between Nu and f versus Re. Also note the three flow regimes: laminar, transition, and turbulent.

Physical and Transport Properties

This subsection summarizes important definitions and data on physical properties of fluids and solid materials. Additional terms and definitions can be found in the glossary.

Density and Specific Gravity

The density of matter is expressed in terms of mass per unit volume. The specific gravity of a material is the ratio of its density to the density of some reference substance. Pure water at 15.5°C is the reference substance for expressing the specific gravities of solids and liquids. Air is

Table 1-1
Summary of Dimensionless Numbers
in Convection Heat Transfer

Dimensionless group	Symbolic Definition	Significance
Nusselt (Nu)	$\dfrac{hL}{k}$	Ratio of temperature gradients
Prandtl (Pr)	$\dfrac{C_p\mu}{k}$	$\dfrac{\text{molecular diffusivity of momentum}}{\text{molecular diffusivity of heat}}$
Peclet (Pe)	$\dfrac{C_p UL}{k}$	$\dfrac{\text{heat transfer by convection}}{\text{heat transfer by conduction}}$
Stanton (St)	$\dfrac{h}{C_p \rho U}$	$\dfrac{\text{wall heat-transfer rate}}{\text{heat transfer by convection}}$
Graetz (Gz)	$Pe\,\dfrac{D}{L}\,\dfrac{\pi}{4}$ or $Pe\,\dfrac{D}{L}$	Same as Peclet number except effect of D/L is accounted for (entrance regions)
Grashot (Gr)	$\dfrac{gL^3}{\nu^2}\beta(T_w - T_\infty)$	$\dfrac{\text{(buoyancy forces) (inertia force)}}{\text{(viscous forces)}^2}$
Viscosity ratio	$\dfrac{\mu_b}{\mu_w}$	$\dfrac{\text{viscosity of fluid at bulk temperature}}{\text{viscosity of fluid at wall temperature}}$

Table 1-2
Summary of Common Symbols

	Symbol	Unit or Definition
General Symbols		
Acceleration	a	m/s^2
of gravity	g	m/s^2
Base of natural logarithms	e	
Coefficient	C	
Difference, finite	Δ	
Efficiency	η	
Energy	E	J, N-m

Table 1-2 Continued

	Symbol	Unit or Definition
General Symbols Continued		
Enthalpy	H	J
Entropy	S	J/K
Force	F	N
Heat	Q	J
Internal energy	U	J
Mass	m	kg
Pressure	p, P	Pa, bar
Shear stress	τ	Pa
Temperature		
Dimensionless	θ	
Absolute	T	K (Kelvin)
In general	T, t	°C
Time		
Dimension of	T	s
In general	t, τ	s, hr
Work	W	J

	Symbol	Unit or Definition
Geometrical Symbols		
Linear dimension		
Breadth	b	m
Diameter	D	m
Distance	x	m
Height above datum plane	z	m
Height equivalent	H	m
Hydraulic radius	r_H	m
Length	L, ℓ	m
Radius	r	m
Thickness	B, δ	
Area		
In general	A	m^2
Cross section	S	m^2
Fraction free cross section	σ	
Projected	A_p	m^2
Surface		
Per unit mass	A_w, s	m^2/kg
Per unit volume	A_s, a	m^2/m^3
Volume		
In general	V	m^3
Humid volume	ν_H	m^3/kg dry air

(Table 1-2 continued on next page)

Table 1-2 Continued

	Symbol	Unit or Definition
Intensive Properties Symbols		
Absorptivity for radiation	α	
Coefficient of expansion		
Linear	α	m/(m-K)
Volumetric	β	m^3/(m-K)
Density	ρ	kg/m^3
Diffusivity		
Molecular, volumetric	D_v	m^3/(s-m), Mm^2/s
Thermal	α	m^2/s
Emissivity ratio for radiation	e	
Humid heat	C_s	J/(kg dry-air-K)
Specific heat	C	J/(kg-K)
At constant pressure	C_p	J/(kg-K)
At constant volume	C_v	J/(kg-K)
Specific heats, ratio of	γ	
Thermal conductivity	k	(J-m)/(s-m^2-K)
Viscosity		
Absolute or coefficient of	μ	Pa-s
Kinematic	ν	m^2/s
Rate Symbols		
Heat transfer rate	q	J/s
Mass rate of flow	W	kg/s, kg/hr
Molal rate of transfer	N	kmol/s
Velocity	u, v	m/s
Volumetric rate of flow	q, Q	m^3/s, m^3/hr
Quantity per unit time, unit area		
Emissive power, total	W	W/m^2
Mass velocity	G	kg/(s-m^2)
Vapor or light phase	G, $\bar{G}$	kg/(s-m^2)
Liquid or heavy phase	L, $\bar{L}$	kg/(s-m^2)
Radiation, intensity of	I	W/m^2
Space velocity, volumetric	Λ	m^3/(s-m^3)
Heat transfer coefficient		
Individual	h	W/(m^2-K)
Overall	U	W/(m^2-K)
Mass transfer coefficient		
Individual	$k_{G,L}$	kmol/(s-m^2)(driving force)
Overall	$K_{G,L}$	p for bar
Stefan-Boltzmann constant	σ	5.6703×10^{-8}W/(m^2-K^4)

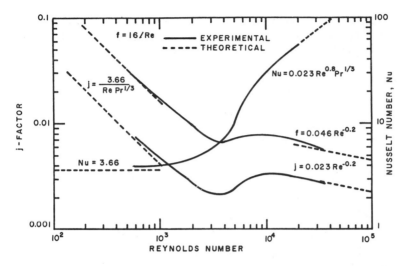

Figure 1-1. Basic heat transfer and flow friction characteristics for flow through a circular tube.

the reference substance for gases; where the specific gravity of a gas is the ratio of its density to the density of air, both at the same temperature and pressure.

Gas Behavior

An equation of state relates pressure, volume, and temperature. For perfect gases, the equation of state is:

$$PV_m = RT \tag{1-25}$$

where P = pressure
  V_m = volume per mole
  T = absolute temperature
  R = gas constant

For real gases, the equation of state can be expressed as:

$$PV_m = ZRT \tag{1-26}$$

where Z is the compressibility factor. The compressibility factor at any condition can be obtained from a plot of the compressibility factor versus reduced pressure at constant values of reduced temperature.

Heat Capacity

This property is defined as the amount of heat required to increase the temperature of a material one degree. If the material is heated at constant pressure, then:

$$C_p = \left(\frac{\partial H}{\partial T}\right)_p \qquad (1\text{-}27)$$

When heating at constant volume:

$$C_\nu = \left(\frac{\partial E}{\partial T}\right)_\nu \qquad (1\text{-}28)$$

where   H = enthalpy
        E = internal energy
        C_p = heat capacity at constant pressure
        $C\nu$ = heat capacity at constant volume

The dimensions of heat capacity are energy per unit mass per unit temperature change.

For perfect gases:

$$C_p - C_\nu = R \qquad (1\text{-}29)$$

For liquids and solids, $C_p \simeq C\nu$. For practical flow calculations, C_p is used even though pressure variations exist in the fluid.

For a temperature change at constant pressure:

$$H = \left(\int_{T_1}^{T_2} C_p dT\right)_p \qquad (1\text{-}30)$$

For both pressure and temperature change, the enthalpy change is computed by:

$$dH = C_p dT + \left[-T\frac{\partial}{\partial T}\left(\frac{1}{\rho}\right) + \frac{1}{\rho}\right]dP \qquad (1\text{-}31)$$

Molecular-Transport Properties

This refers to those properties concerned with the rate of momentum, heat, and mass transfer by molecular motion. The rate is proportional to the potential gradient, and the constant of proportionality is a physical property of the material.

For momentum transfer:

$$\frac{F}{A} = -\mu \frac{du}{dy} \qquad (1\text{-}32)$$

$$\frac{\text{momentum transfer}}{(\text{unit area})(\text{unit time})} = (\text{viscosity})(\text{velocity gradient})$$

For heat transfer:

$$\frac{q}{A_q} = -k \frac{dT}{dy} \qquad (1\text{-}33)$$

$$\frac{\text{heat transfer}}{(\text{unit area})(\text{unit time})} = (\text{thermal conductivity})(\text{temperature gradient})$$

For mass transfer:

$$\frac{N_m}{A_{Nm}} = -D \frac{dC_m}{dy} \qquad (1\text{-}34)$$

$$\frac{\text{mass transfer}}{(\text{unit area})(\text{unit time})} = (\text{diffusion coefficient})(\text{gradient})$$

Thermal Conductivity

This property provides a measure of the ability of a material to transfer heat by molecular conduction. Equation 1-33 is the differential equation for the one-dimensional conduction of heat in a substance.

Note
q = rate of heat flow per unit time
A_q = area of flow
dT/dy = temperature gradient in material
k = thermal conductivity of substance

The units of k are in energy transfer per unit cross-sectional area per unit temperature gradient. An approximate correlation for predicting thermal conductivities of gases is:[1]

$$k = \mu\left(C_p + \frac{5R}{4M}\right) \tag{1-35}$$

where k = Btu/(hr)(ft^2)(°F)/ft
        μ = lb$_m$/(ft)(hr)
        C_p = Btu/(lb$_m$)(°F)
        R = 1.987 Btu/(lb-mole)(°R)
        M = molecular weight

Thermal conductivity of liquids can be estimated from:[2]

$$k = 22.9C_p\left(\frac{\rho}{M}\right)^{4/3}\frac{T_B}{\Delta H_\nu} \tag{1-36}$$

where C_p = Btu/(lb$_m$)(°F)
        ρ = g/cm^3
        M = molecular weight
        ΔH_ν = latent heat of vaporization at T_B (Btu/lb$_m$)
        T_B = normal boiling point (°F)

Diffusion Coefficient

This property provides a measure of the rate of molecular diffusion (mass transfer) of one or more components in a mixture under the influence of a concentration gradient. Diffusion takes place in the direction of decreasing concentration. Equation 1-34 is the differential equation for one-dimensional diffusion where N_m is the molal rate of diffusion; A_{Nm} is area; dC_m/dy is the concentration gradient of the diffusing substance; D is the diffusion coefficient.

For gases, the following correlation can be used to predict diffusion coefficients:[3]

$$D = 0.0043\frac{T^{3/2}}{P\left(V_{m1}^{1/3} + V_{m2}^{1/3}\right)}\sqrt{\frac{1}{M_1} + \frac{1}{M_2}} \tag{1-37}$$

where V_{m1}, V_{m2} = respective molecular volumes of gases 1 and 2 (cc)

M_1, M_2 = molecular weights of gases

$\quad\quad T$ = absolute temperature (°K)

$\quad\quad D$ = diffusivity (cm²/s)

Various physical properties data for liquids, solids, and gases are given in Table 1-3 through 1-9. Table 1-3 gives specific heats of miscellaneous materials. Table 1-4 gives additional mean specific heats of different substances. Table 1-5 gives heat capacity data for pure water. Table 1-6 gives data on the coefficients of thermal expansion of different gases. Table 1-7 gives general physical properties for selected solid materials. Table 1-8 gives general physical properties of selected gases and liquids. Finally, Table 1-9 provides abstracts of the ASME steam tables.

Table 1-3
Specific Heats of Miscellaneous Materials

Material	Specific Heat (g-cal per g per °C)
Aluminum	0.2 (100°C); 0.274 (1500°C)
Asbestos	0.25
Asphalt	0.22
Carbon	0.168 (26 ~ 76°C); 0.314 (40 ~ 892°C); 0.387 (56 ~ 1450°C)
Cellulose	0.32
Clay	0.224
Coal	0.26 ~ 0.38
Coal tars	0.35 (40°C); 0.45 (200°C)
Coke	0.265 (21 ~ 400°C); 0.359 (21 ~ 800°C); 0.403 (21 ~ 1300°C)
Gasoline	0.53
Glass (pyrex)	0.20
(silicate)	0.188 ~ 0.204 (0 ~ 100°C); 0.24 ~ 0.26 (0 ~ 700°C)
Graphite	0.165 (26 ~ 76°C); 0.390 (56 ~ 1450°C)
Kerosene	0.47
Sand	0.191
Silica	0.316
Turpentine	0.42 (18°C)
Wood	0.45 ~ 0.65

Table 1-4
Mean Specific Heats of Different Substances
(Btu per lb per °F) ($\bar{C}_p$ = H/t − 60)

t (°F)	O_2	H_2	H_2O	N_2	CO	CO_2
100	0.2188	3.420	0.4448	0.2482	0.4285	0.2022
200	0.2203	3.434	0.4472	0.2485	0.2488	0.2086
400	0.2240	3.448	0.4529	0.2493	0.2501	0.2201
600	0.2279	3.455	0.4597	0.2509	0.2522	0.2301
800	0.2318	3.462	0.4674	0.2531	0.2549	0.2388
1000	0.2355	3.470	0.4757	0.2558	0.2580	0.2465
1200	0.2390	3.480	0.4844	0.2586	0.2611	0.2533
1400	0.2420	3.494	0.4932	0.2614	0.2642	0.2593
1600	0.2448	3.510	0.5021	0.2642	0.2672	0.2646
1800	0.2473	3.528	0.5111	0.2669	0.2700	0.2694
2000	0.2495	3.549	0.5201	0.2695	0.2726	0.2737
2200	0.2517	3.572	0.5289	0.2719	0.2751	0.2776
2400	0.2536	3.596	0.5375	0.2742	0.2774	0.2813
2600	0.2554	3.620	0.5456	0.2764	0.2794	0.2845
2800	0.2570	3.644	0.5536	0.2784	0.2814	0.2875
3000	0.2585	3.668	0.5614	0.2802	0.2831	0.2902
3200	0.2600	3.692	0.5688	0.2819	0.2848	0.2927

Sample Calculation 1-1. Estimate the gas conductivity for the synthesis of $CO + 2H_2$ at t = 180°C and ambient pressure.

Solution. Both components are diatomic gases, and have approximately the same molar heat capacities (C_p = 7, C_v = 5). This gives a heat capacity ratio of C_p/C_v = 1.4. Use the following formula:

$$Pr = \frac{4}{q - \dfrac{5}{C_p/C_v}}$$

$$Pr = \frac{4}{9 - \dfrac{5}{1.4}} = 0.74$$

From literature tables on the viscosity for hydrogen, μ_0 = 0.0084 cP and carbon monoxide, μ_0 = 0.0166 cP at 0°C.

Compute the heat capacity of these gases at 0°C from the expression for the Prandtl number:

$$Pr = \frac{C_p \mu g}{k}$$

First determine the heat capacities of both gases according to the following parameters: their molecular heat capacity, $C_p = 7$ kcal/kg-

Table 1-5
Heat Capacity of Water
(Data for Air-Free Water at Constant Pressure of 1 Atm)

Temperature (°C)	(°F)	Heat Capacity (cal/g)	Temperature (°C)	(°F)	Heat Capcity (cal/g)
0	32.00	1.00803	52	125.60	0.99935
2	35.60	1.00636	54	129.20	0.99950
4	39.20	1.00495	56	132.80	0.99969
6	42.80	1.00378	58	136.40	0.99988
8	46.40	1.00277	60	140.00	1.00007
10	50.00	1.00194	62	143.60	1.00029
12	53.60	1.00124	64	147.20	1.00053
14	57.20	1.00067	66	150.80	1.00079
16	60.80	1.00019	68	154.40	1.00105
18	64.40	0.99978	70	158.00	1.00131
20	68.00	0.99947	72	161.60	1.00160
22	71.60	0.99921	74	165.20	1.00191
24	75.20	0.99902	76	168.80	1.00225
26	78.80	0.99885	78	172.40	1.00258
28	82.40	0.99873	80	176.00	1.00294
30	86.00	0.99866	82	179.60	1.00332
32	89.60	0.99861	84	183.20	1.00373
34	93.20	0.99859	86	186.80	1.00414
36	96.80	0.99861	88	190.40	1.00457
38	100.40	0.99864	90	194.00	1.00502
40	104.00	0.99869	92	197.60	1.00550
42	107.60	0.99876	94	201.20	1.00600
44	111.20	0.99883	96	204.80	1.00653
46	114.80	0.99895	98	208.40	1.00705
48	118.40	0.99907	100	212.00	1.00763
50	122.00	0.99919			

Source: Osborne, Stimson and Ginnings, J. Research Natl. Bur. Standards, *23:197 (1939).*
Note: 1 calorie = 4.1833 NBS int. J. (National Bureau of Standards international joule).

Table 1-6
Coefficients of Thermal Expansion of Different Gases

Constant Volume Coefficient for Temperatures 0° ~ 100°C $\alpha_p = \dfrac{1}{P_o}\left(\dfrac{dP}{dt}\right)_v$			Constant Pressure Coefficient for Temperatures 0 ~ 100°C $\alpha_v = \dfrac{1}{v_o}\left(\dfrac{dv}{dt}\right)_p$		
Material	Initial Pressure (mm Hg)	$10^6\alpha_p$	Material	Pressure (mm Hg)	$10^6\alpha_v$
Air	760	3,671.6	Air	760	3,671.1
	1,000	3,675		1,000	3,674
Ammonia	760	3,767.8	Ammonia	760	3,790
Carbon dioxide	760	3,711	Carbon dioxide	760	3,725
	1,000	3,726	Carbon monoxide	760	3,672
Carbon monoxide	760	3,673	Ethylene	760	3,735
Ethylene	760	3,722	Helium	760	3,659.1
Helium	760	3,661.3		994	3,657.9
	1,000	3,660.7	Hydrogen	760	3,660.3
Hydrogen	760	3,662.7		1,095	3,659.0
	1,000	3,662.6	Methane	760	3,682
Methane	760	3,679	Nitrogen	760	3,671
Nitrogen	760	3,672		994	3,673.4
	994	3,674	Oxygen	760	3,674
Oxygen	760	3,673.5		1,000	3,676.3
	1,000	3,675.7	Sulfur dioxide	760	3,880
Sulfur dioxide	760	3,840			

Source: Coppock, Phil. Mag., 7(19):446 (1935).

mole; the molecular weight of hydrogen, $M = 2$; the molecular weight of carbon monoxide, $M = 28$. The specific heat capacity is:

$$H_2 : C_{p1} = \frac{7}{2} = 3.5 \text{ kcal/kg-°C}$$

$$CO : C_{p2} = \frac{7}{28} = 0.25 \text{ kcal/kg-°C}$$

Dividing viscosity μ in cP by 10^3 we obtain the product μg in technical units (kgf,m,s).

Further, we determine the gas conductivity from the Prandtl number:

$$H_2 = k_0 = \frac{3.5(0.0084 \times 10^{-3})}{0.74} = 3.98 \times 10^{-5} \frac{kcal}{m\text{-}s\text{-}°C}$$

$$CO = k_0 = \frac{0.25(0.016 \times 10^{-3})}{0.74} = 5.60 \times 10^{-6} \frac{kcal}{m\text{-}s\text{-}°C}$$

Multiplying these values by 3,600 we obtain conductivities of both components in technical units:

$$H_2 : k_0 = 3.98 \times 10^{-5} \times 3,600 = 1.44 \times 10^{-1} \frac{kcal}{m\text{-}hr\text{-}°C}$$

{text continued on page 26}

Table 1-7
Physical Properties of Selected Solid Materials

Material	Density ρ (lb_m/ft^3) (68°F)	Specific Heat c_p (Btu/lb_m °F) (68°F)	Diffusivity α (ft^2/hr) (68°F)	Thermal Conductivity k (Btu/hr ft °F) (68°F)	(212°F)	(572°F)
Aluminum	168.6	0.224	3.55	132	132	133
Brass (70% Cu, 30% Zn)	532	0.091	1.27	61.8	73.9	85.3
Copper	555	0.092	3.98	223	219	213
Gold	1,206	0.031	4.52	169	170	172
Iron	492	0.122	0.83	42.3	39.0	31.6
Iron, cast	455	0.100	0.65	29.6	26.8	
Lead	708	0.030	0.80	20.3	19.3	17.2
Magnesium	109	0.248	3.68	99.5	96.8	91.4
Nickel	556	0.111	0.87	53.7	47.7	36.9
Platinum	1,340	0.032	0.09	40.5	41.9	43.5
Silver	656	0.057	6.42	240	237	209
Stainless steel	488	0.110	0.17	9.4	10.0	13
Steel, mild (1% C)	488	0.113	0.45	24.8	24.8	22.9
Tungsten	1,206	0.032	2.44	94	87	77
Zinc	446	0.094	1.55	65	63	58

Table 1-8

Physical Properties of Selected Liquids and Gases*

T (°F)	ρ (lb$_m$/ft³)	c_p (Btu/lb$_m$ °F)	μ (lb$_m$/ft s)	ν (ft²/s)	k (Btu/hr ft °F)	α (ft²/hr)	Pr	β (1/°F)	$g\beta\rho^2/\mu^2$ (1/°F-ft³)
Air (gas-1 atm pressure)									
0	0.0862	0.240	1.09(×10⁵)	0.126(×10³)	0.0132(×10³)	0.639	0.721	2.18(×10³)	4.39(×10⁶)
60	0.0764	0.240	1.21	0.159	0.0146	0.798	0.711	1.92	2.48
100	0.0710	0.240	1.28	0.181	0.0156	0.919	0.703	1.79	1.76
200	0.0602	0.241	1.45	0.241	0.0179	1.24	0.694	1.52	0.840
300	0.0523	0.243	1.60	0.306	0.0203	1.60	0.686	1.32	0.454
400	0.0462	0.245	1.74	0.377	0.0225	2.00	0.681	1.16	0.264
500	0.0413	0.247	1.87	0.453	0.0246	2.41	0.680	1.04	0.163
600	0.0374	0.251	2.00	0.535	0.0270	2.88	0.680	0.944	79.4(×10³)
800	0.0315	0.257	2.24	0.711	0.0303	3.75	0.684	0.794	50.6
1,000	0.0272	0.263	2.46	0.906	0.0337	4.72	0.689	0.685	27.0
Carbon dioxide (gas-1 atm pressure)									
0	0.132	0.193	0.865(×10⁵)	0.0655(×10³)	0.00760(×10³)	0.298	0.792	2.18(×10³)	16.3(×10⁶)
60	0.117	0.202	0.965	0.0829	0.00910	0.387	0.773	1.92	9.00
100	0.108	0.207	1.03	0.0953	0.0102	0.455	0.758	1.79	6.33
200	0.092	0.219	1.20	0.131	0.0130	0.646	0.730	1.52	2.86
300	0.0800	0.230	1.36	0.171	0.0160	0.878	0.704	1.32	1.45
400	0.0740	0.239	1.45	0.196	0.0180	1.02	0.695	1.16	1.11
500	0.0630	0.248	1.65	0.263	0.0210	1.36	0.700	1.04	0.485
600	0.0570	0.256	1.78	0.312	0.0235	1.61	0.700	0.944	0.310
800	0.0480	0.269	2.02	0.420	0.0278	2.15	0.702	0.794	0.143
1,000	0.0416	0.280	2.25	0.540	0.0324	2.78	0.703	0.685	75.3(×10³)

*($\times 10^n$ = actual values have been multiplied by powers of 10).

Table 1-8 Continued

T (°F)	ρ (lb$_m$/ft³)	c_p (Btu/lb$_m$ °F)	μ (lb$_m$/ft s)	ν (ft²/s)	k (Btu/hr ft °F)	α (ft²/hr)	Pr	β (1/°F)	$g\beta\rho^2/\mu^2$ (1/°F·ft³)
Sulfur dioxide (gas-1 atm pressure)									
0	0.195	0.142	0.700(×10⁵)	3.59(×10³)	0.00460	0.166	0.778	2.03(×10³)	50.6(×10⁶)
100	0.161	0.149	0.890	5.52	0.00560	0.233	0.854	1.79	19.0
200	0.136	0.157	1.05	7.74	0.00670	0.313	0.883	1.52	8.25
300	0.118	0.164	1.20	10.2	0.00790	0.407	0.898	1.32	4.12
400	0.104	0.170	1.35	13.0	0.00920	0.520	0.898	1.16	2.24
500	0.0935	0.176	1.50	16.0	0.00990	0.601	0.958	1.04	1.30
600	0.0846	0.180	1.65	19.5	0.0108	0.711	0.987	0.994	0.795
Ammonia									
-60	43.9	1.07	20.6(×10⁵)	0.471(×10⁵)	0.316	6.74(×10³)	2.52	0.94(×10³)	132(×10⁷)
-30	42.7	1.07	18.2	0.426	0.317	6.93	2.22	1.02	265
0	41.3	1.08	16.9	0.409	0.315	7.06	2.08	1.1	467
30	40.0	1.11	16.2	0.402	0.312	7.05	2.05	1.19	757
60	38.5	1.14	15.0	0.391	0.304	6.92	2.03	1.3	1,130
80	37.5	1.16	14.2	0.379	0.296	6.79	2.01	1.4	1,650
100	36.4	1.19	13.5	0.368	0.287	6.62	2.00	1.5	2,200
Water									
32	62.4	1.01	1.20(×10³)	1.93(×10⁵)	0.319	5.06(×10³)	13.7	-0.350(×10⁴)	
60	62.3	1.00	0.760	1.22	0.340	5.45	8.07	0.800	17.2(×10⁻⁶)
80	62.2	0.999	0.578	0.929	0.353	5.67	5.89	1.30	48.3
100	62.1	0.999	0.458	0.736	0.364	5.87	4.51	1.80	107
200	60.1	1.01	0.206	0.342	0.392	6.46	1.91	3.70	1,010
300	57.3	1.03	0.130	0.227	0.395	6.70	1.22	5.60	3,350
400	53.6	1.08	0.0930	0.174	0.382	6.58	0.950	7.80	8,350
500	49.0	1.19	0.0700	0.143	0.349	5.98	0.859	11.0	17,350
600	42.4	1.51	0.0579	0.137	0.293	4.58	1.07	17.5	30,300

(Table 1-8 continued on next page)

Table 1-8 Continued

T (°F)	ρ (lb$_m$/ft^3)	c_p (Btu/lb$_m$ °F)	μ (lb$_m$/ft s)	ν (ft^2/s)	k (Btu/hr ft °F)	α (ft^2/hr)	Pr	β (1/°F)	$g\beta\rho^2/\mu^2$ (1/°F-ft^3)
					Kerosene				
30	48.8	0.456	800($\times 10^5$)	16.4($\times 10^5$)	0.0809	3.63($\times 10^3$)	163		
60	48.1	0.474	600	12.5	0.0805	3.53	127	0.58($\times 10^3$)	120($\times 10^{-4}$)
80	47.6	0.491	490	10.3	0.0800	3.42	108	0.48	146
100	47.2	0.505	420	8.90	0.0797	3.35	95.7	0.47	192
					Glycerin				
30	79.7	0.540	7.2	9.03($\times 10^2$)	0.168	3.91($\times 10^3$)	832($\times 10^2$)		
60	79.1	0.563	1.4	1.77	0.167	3.75	170		
80	78.7	0.580	0.6	0.762	0.166	3.64	75.3	0.30($\times 10^3$)	166
100	78.2	0.598	0.1	0.128	0.165	3.53	13.1		
					n-Butyl Alcohol				
60	50.5	0.55	225($\times 10^5$)	4.46($\times 10^5$)	0.100	3.59($\times 10^3$)	44.6	0.25($\times 10^3$)	6.23($\times 10^{-6}$)
80	50.0	0.58	180	3.60	0.099	3.41	38.0	0.43	2.02
100	49.6	0.61	130	2.62	0.098	3.25	29.1		
150	48.5	0.68	68	1.41	0.098	2.97	17.1		
					Benzene				
60	55.2	0.395	44.5($\times 10^5$)	0.806($\times 10^5$)	0.0856	3.93($\times 10^3$)	7.39	7.5($\times 10^4$)	498($\times 10^{-6}$)
80	54.6	0.410	38	0.695	0.0836	3.73	6.70	7.2	609
100	53.6	0.420	33	0.615	0.0814	3.61	6.13		
200	49.9	0.480	19.4	0.390	0.0711	2.97	4.73		

Table 1-9
Properties of Saturated Steam

Abs. Press (lb/sq in.) p	Temp. (°F) t	Specific Volume Sat. Liquid v_f	Specific Volume Evap v_{gf}	Specific Volume Sat. Vapor v_g	Enthalpy Sat. Liquid h_f	Enthalpy Evap h_{fg}	Enthalpy Sat. Vapor h_g	Entropy Sat. Liquid s_f	Entropy Evap s_{fg}	Entropy Sat. Vapor s_g
0.08865	32.018	0.016022	3302.4	3302.4	0.0003	1075.5	1075.5	0.0000	2.1872	2.1872
0.25	59.323	0.016032	1235.5	1235.5	27.382	1060.1	1087.4	0.0542	2.0425	2.0967
0.50	79.586	0.016071	641.5	641.5	47.623	1048.6	1096.3	0.0925	1.9446	2.0370
1.0	101.74	0.016136	333.59	333.60	69.73	1036.1	1105.8	0.1326	1.8455	1.9781
5.0	162.24	0.016407	73.515	73.532	130.20	1000.9	1131.1	0.2349	1.6094	1.8443
10.0	193.21	0.016592	38.404	38.420	161.26	982.1	1143.3	0.2836	1.5043	1.7879
14.696	212.00	0.016719	26.782	26.799	180.17	970.3	1150.5	0.3121	1.4447	1.7568
15.0	213.03	0.016726	26.274	26.290	181.21	969.7	1150.9	0.3137	1.4415	1.7552
20.0	227.96	0.016834	20.070	20.087	196.27	960.1	1156.3	0.3358	1.3962	1.7320
30.0	250.34	0.017009	13.7266	13.7436	218.9	945.2	1164.1	0.3682	1.3313	1.6995
40.0	267.25	0.017151	10.4794	10.4965	236.1	933.6	1169.8	0.3921	1.2844	1.6765
50.0	281.02	0.017274	8.4967	8.5140	250.2	932.9	1174.1	0.4112	1.2474	1.6586
60.0	292.71	0.017383	7.1562	7.1736	262.2	915.4	1177.6	0.4273	1.2167	1.6440
70.0	302.93	0.017482	6.1875	6.2050	272.7	907.8	1180.6	0.4411	1.1905	1.6316
80.0	312.04	0.017573	5.4536	5.4711	282.1	900.9	1183.1	0.4534	1.1675	1.6208
90.0	320.28	0.017659	4.8779	4.8953	290.7	894.6	1185.3	0.4643	1.1470	1.6113
100.0	327.82	0.017740	4.4133	4.4310	298.5	888.6	1187.2	0.4743	1.1284	1.6027
120.0	341.27	0.01789	3.7097	3.7275	312.6	877.8	1190.4	0.4919	1.0960	1.5879
140.0	353.04	0.01803	3.2010	3.2190	325.0	868.0	1193.0	0.5071	1.0681	1.5752

(Table 1-9 continued on next page)

Table 1-9 Continued

Abs. Press (lb/sq in.) p	Temp. (°F) t	Specific Volume			Enthalpy			Entropy		
		Sat. Liquid v_f	Evap v_{gf}	Sat. Vapor v_g	Sat. Liquid h_f	Evap h_{fg}	Sat. Vapor h_g	Sat. Liquid s_f	Evap s_{fg}	Sat. Vapor s_g
160.0	363.55	0.01815	2.8155	2.8336	336.1	859.0	1195.1	0.5206	1.0435	1.5641
180.0	373.08	0.01827	2.5129	2.5312	346.2	850.7	1196.9	0.5328	1.0215	1.5543
200.0	381.80	0.01839	2.2689	2.2873	355.5	842.8	1198.3	0.5438	1.0016	1.5454
220.0	389.88	0.01850	2.06779	2.08629	364.2	835.4	1199.6	0.5540	0.9834	1.5374
240.0	397.39	0.01860	1.89909	1.91769	372.3	828.4	1200.6	0.5634	0.9665	1.5299
260.0	404.44	0.01870	1.75548	1.77418	379.9	821.6	1201.5	0.5722	0.9508	1.5230
280.0	411.07	0.01880	1.63169	1.65049	387.1	815.1	1202.3	0.5805	0.9361	1.5166
300.0	417.35	0.01889	1.52384	1.54274	394.0	808.9	1202.9	0.5882	0.9223	1.5105
350.0	431.73	0.01912	1.30642	1.32554	409.8	794.2	1204.0	0.6059	0.8909	1.4968
400.0	444.60	0.01934	1.14162	1.16095	424.2	780.4	1204.6	0.6217	0.8630	1.4847
450.0	456.28	0.01954	1.01224	1.03179	437.3	767.5	1204.8	0.6360	0.8378	1.4738
500.0	467.01	0.01975	0.90787	0.92762	449.5	755.1	1204.7	0.6490	0.8148	1.4639
550.0	476.94	0.01994	0.82183	0.84177	460.9	743.3	1204.3	0.6611	0.7936	1.4547
600.0	486.20	0.02013	0.74962	0.76975	471.7	732.0	1203.7	0.6723	0.7738	1.4461
650.0	494.89	0.02032	0.68811	0.70843	481.9	720.9	1202.8	0.6828	0.7552	1.4381
700.0	503.08	0.02050	0.63505	0.65556	491.6	710.2	1201.8	0.6928	0.7377	1.4304
800.0	518.21	0.02087	0.54809	0.56896	509.8	689.6	1199.4	0.7111	0.7051	1.4163
900.0	531.95	0.02123	0.47968	0.50091	526.7	669.7	1196.4	0.7279	0.6753	1.4032
1000.0	544.58	0.02159	0.42436	0.44596	542.6	650.4	1192.9	0.7434	0.6476	1.3910
1200.0	567.19	0.02232	0.34013	0.36245	571.9	613.0	1184.8	0.7714	0.5969	1.3683

Table 1-9 Continued

Abs. Press (lb/sq in.) p	Temp. (°F) t	Specific Volume			Enthalpy			Entropy		
		Sat. Liquid v_f	Evap v_{gf}	Sat. Vapor v_g	Sat. Liquid h_f	Evap h_{fg}	Sat. Vapor h_g	Sat. Liquid s_f	Evap s_{fg}	Sat. Vapor s_g
1400.0	587.07	0.02307	0.27871	0.30178	598.8	576.5	1175.3	0.7966	0.5507	1.3474
1600.0	604.87	0.02387	0.23159	0.25545	624.2	540.3	1164.5	0.8199	0.5076	1.3274
1800.0	621.02	0.02472	0.19390	0.21861	648.5	503.8	1152.3	0.8417	0.4662	1.3079
2000.0	635.80	0.02565	0.16266	0.18831	672.1	466.2	1138.3	0.8625	0.4256	1.2881
2200.0	649.45	0.02669	0.13603	0.16272	695.5	426.7	1122.2	0.8828	0.3848	1.2676
2400.0	662.11	0.02790	0.11287	0.14076	719.0	384.8	1103.7	0.9031	0.3430	1.2460
2600.0	673.91	0.02938	0.09172	0.12110	744.5	337.6	1082.0	0.9247	0.2977	1.2225
2800.0	684.96	0.03134	0.07171	0.10305	770.7	285.1	1055.8	0.9468	0.2491	1.1958
3000.0	695.33	0.03428	0.05073	0.08500	801.8	218.4	1020.3	0.9728	0.1891	1.1619
3200.0	705.08	0.04472	0.01191	0.05663	875.5	56.1	931.6	1.0351	0.0482	1.0832
3208.2*	705.47	0.05078	0.00000	0.05078	906.0	0.0	906.0	1.0612	0.0000	1.0612

* Critical pressure.
Courtesy of the American Society of Mechanical Engineers

$$CO : k_0 = 5.60 \times 10^{-6} \times 3{,}600 = 2.02 \times 10^{-2}$$

Use the following equation:

$$k = k_0 \left(\frac{273 + S}{T + S}\right)\left(\frac{T}{273}\right)^{3/2}$$

where k_0 = conductivity of gas at $0°C$
T = absolute temperature of gas
S = constant; for hydrogen $S = 83$, for carbon monoxide $S = 118$
$T = 180 + 273 = 453°K$

The conductivity of both gases at $180°C$ is:

$$H_2 : 1.44 \times 10^{-1}\left(\frac{273 + 83}{453 + 83}\right)\left(\frac{453}{273}\right)^{3/2} = 0.205\frac{kcal}{m\text{-}hr\text{-}°C}$$

$$CO : 2.02 \times 10^{-2}\left(\frac{273 + 118}{453 + 118}\right)\left(\frac{453}{273}\right)^{3/2} = 0.0296\frac{kcal}{m\text{-}hr\text{-}°C}$$

The conductivity of the mixture is determined from the following equation:

$$k_m = \frac{k_1}{1 + A_{1,2}\dfrac{x_2}{x_1}} + \frac{k_2}{1 + A_{2,1}\dfrac{x_1}{x_2}} \tag{1-38}$$

where k_1, k_2 = conductivities of components;
x_1, x_2 = volume fraction of component
$A_{1,2}, A_{2,1}$ = constants from the following formula

$$A_{1,2} = \frac{1}{4}\left\{1 + \left[\frac{\mu_1}{\mu_2}\left(\frac{M_2}{M_1}\right)^{3/4}\left(\frac{1 + S_1/T}{1 + S_2/T}\right)\right]^{1/2}\right\}^2\frac{\left(1 + \dfrac{S_{1,2}}{T}\right)}{\left(1 + \dfrac{S_1}{T}\right)}$$

Changing indexes 1 and 2 we obtain a similar expression for constant $A_{2,1}$; S_1 and S_2 = constants.

$$S_{1,2} = \sqrt{S_1 S_2} = \sqrt{83 \times 118} = 99$$

$$\frac{\mu_1}{\mu_2} = \frac{k_1}{k_2} \frac{C_{P2}}{C_{P1}} \frac{(q - 5(C_P/C_\nu)_2)}{(q - 5(C_P/C_\nu)_1)}$$

$$= \frac{0.205}{0.0296} \times \frac{0.25}{3.5} \left(\frac{q - 5 \times 1.4}{q - 5 \times 1.4} \right) = 0.495$$

$$\frac{\mu_2}{\mu_1} = 2.02$$

$$A_{1,2} = \frac{1}{4} \left\{ 1 + \left[0.495 \left(\frac{28}{2} \right)^{3/2} \right. \right.$$
$$\left. \left. \left(\frac{1 + 83/453}{1 + 118/453} \right) \right]^{1/2} \right\}^2 \left(\frac{1 + 99/453}{1 + 83/453} \right) = 2.3$$

$$A_{2,1} = \frac{1}{4} \left\{ 1 + \left[2.02 \left(\frac{2}{28} \right)^{3/4} \right. \right.$$
$$\left. \left. \left(\frac{1 + 118/453}{1 + 83/453} \right) \right]^{1/2} \right\}^2 \left(\frac{1 + 99/453}{1 + 118/453} \right) = 0.57$$

The gas composition corresponds to

$$CO + 2H_2, \quad \frac{x_1}{x_2} = 2 \quad \text{and} \quad \frac{x_2}{x} = 0.5.$$

Substituting these values in Equation 1-38 the conductivity of the mixture at 180°C is:

$$k_m = \frac{0.205}{1 + 2.3 \times 0.5} + \frac{0.0296}{1 + 0.570 \times 2} = 0.109 \frac{kcal}{m\text{-}hr\text{-}°C}$$

References

1. Euchen, A., *Physik. Z.*, 12:1,101 (1911).
2. Palmer, G., *Ind. Eng. Chem.*, 40:89 (1948).
3. Gilliland, E. R., *Ind. Eng. Chem.*, 26:681 (1934).

2

CONDUCTION

Conductivity

The conductivity of metals in general decreases with increasing temperature. Impurities in metals result in a marked reduction in their conductivity. For example, the conductivity of pure copper is approximately 340 kcal/m-hr-°C; however, with trace quantities of arsenic k = 122 kcal/m-hr-°C. For iron with 0.1% carbon k = 45 kcal/m-hr-°C, with 1% C k = 40 kcal/m-hr-°C and with 1.5% C k = 31 kcal/m-hr-°C.

The conductivity of solid nonmetals (i.e., construction and insulating materials) generally increases with temperature (one exception is magnesite brick). Figure 2-1 gives a plot of conductivity versus temperature for different insulating and refractory materials.

Conductivity depends on several factors; key ones being the material structure, porosity, and humidity. Materials with conductivity less than 0.2 are used for thermal insulation.

Table 2-1 lists typical values of k for insulating and building materials at different temperatures.

General Notes on Conduction

The differential equation describing unsteady-state heat conduction is know as Fourier's law:

$$\frac{\partial t}{\partial \tau} = \frac{k}{C\rho} \left(\frac{\partial^2 t}{\partial x^2} + \frac{\partial^2 t}{\partial y^2} + \frac{\partial^2 t}{\partial z^2} \right) \tag{2-1}$$

Note $\alpha = k/C\rho \equiv$ thermal diffusivity where units are:

$$\frac{k}{C\rho} = \left(\frac{kcal/m\text{-}hr\text{-}°C}{kcal/kg\text{-}°C \times kg/m^3} \right) = \left(\frac{m^2}{hr} \right)$$

28

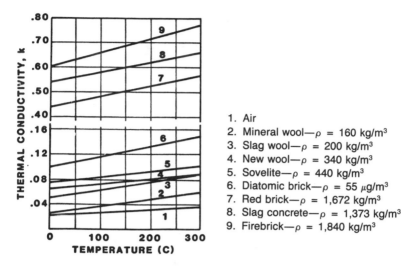

1. Air
2. Mineral wool—ρ = 160 kg/m³
3. Slag wool—ρ = 200 kg/m³
4. New wool—ρ = 340 kg/m³
5. Sovelite—ρ = 440 kg/m³
6. Diatomic brick—ρ = 55 μg/m³
7. Red brick—ρ = 1,672 kg/m³
8. Slag concrete—ρ = 1,373 kg/m³
9. Firebrick—ρ = 1,840 kg/m³

Figure 2-1. k = f(t) for different insulating and refractory materials.

The polynomial inside the parentheses of Equation 2-1 is the Laplace operator ($\nabla^2 t$); hence:

$$\frac{\partial t}{d\tau} = \alpha \nabla^2 t \tag{2-2}$$

There are many problems in heat conduction that can be analyzed more conveniently in cylindrical and spherical coordinate systems (see Figure 2-2).

The heat conduction equation in cylindrical coordinates is:

$$\frac{\partial^2 t}{\partial r^2} + \frac{1}{r}\frac{\partial t}{\partial r} + \frac{1}{r^2}\frac{\partial^2 t}{\partial \phi^2} + \frac{\partial^2 t}{\partial z^2} = \frac{1}{\alpha}\frac{\partial t}{\partial \tau} \tag{2-3}$$

and in spherical coordinates:

$$\frac{1}{r^2}\frac{\partial}{\partial r}\left(r^2 \frac{\partial t}{\partial r}\right) + \frac{1}{r^2 \sin \phi}\frac{\partial}{\partial \phi}\left(\sin \phi \frac{\partial t}{\partial \phi}\right)$$

$$+ \frac{1}{r^2 \sin^2 \phi}\frac{\partial^2 t}{\partial \phi^2} = \frac{1}{\alpha}\frac{\partial t}{\partial \tau} \tag{2-4}$$

Table 2-1
Thermal Conductivities of Typical Insulating
and Building Materials

Material	Temperature (°C)	Thermal Conductivity (kcal/m-hr-°C)
Asbestos	0	0.13
Glass wool	0	0.03
	300	0.09
Cork in slabs	0	0.03
	50	0.04
Magnesia	50	0.05
Slag wool	0	0.05
	200	0.07
Common brick	25	0.34
Porcelain	95	0.89
	1,100	1.70
Concrete	0	1.2
Fresh earth	0	2.0
Glass	15	0.60
Wood (pine):		
Perpendicular fibers	15	0.13
Parallel fibers	20	0.30
Burnt clay	15	0.80
Carborundum	600	16.0

Fourier's equation enables analysis of problems associated with the propagation of heat through a body via conduction in either steady-state or unsteady-state systems.

In heat transfer equipment, heating surfaces are flat, cylindrical, or spherical in shape. Therefore, the solutions of problems of heat transfer by conduction through such geometries are of practical importance.

Heat Conduction Through a Flat Wall

This system is illustrated in Figure 2-3. For steady-state heat conduction the temperature at different points in the wall does not change with time, i.e., $dt/d\tau = 0$; and the temperature distribution is one dimensional:

$$\frac{d^2t}{\partial y^2} = \frac{\partial^2t}{\partial z^2} = 0 \qquad (2\text{-}5)$$

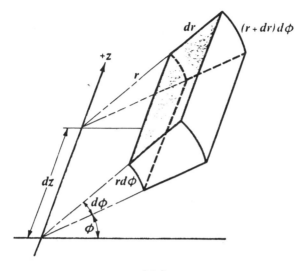

(A)

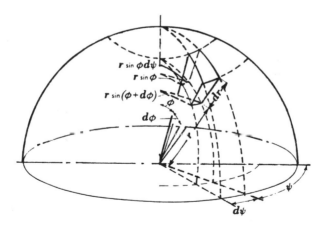

(B)

Figure 2-2. (A) Cylindrical coordinates; (B) spherical coordinates.

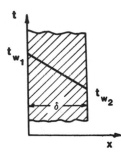

Figure 2-3. Heat conduction through a flat wall.

For this system, Equation 2-1 takes the form:

$$\frac{d^2t}{dx^2} = 0 \tag{2-6}$$

and integrating:

$$\frac{dt}{dx} = C_1$$

$$t = C_1 x + C_2$$

The constants of integration are determined from the boundary conditions ($x = 0$ and $x = \delta$):

$$C_2 = t_{w1}$$

$$C_1 = \frac{dt}{dx} = \frac{t_{w2} - t_{w1}}{\delta}$$

or

$$t = \frac{t_{w2} - t_{w1}}{\delta}x + t_{w2} \tag{2-7}$$

i.e., at steady-state conduction, the temperature along the thickness of a flat wall changes linearly and the temperature gradient remains constant.

The basic law describing one-dimensional heat conduction for the flat wall case is:

$$dQ = k\frac{t_{w1} - t_{w2}}{\delta} \, dAd\tau$$

or

$$Q = \frac{k}{\delta}(t_{w1} - t_{w2})A\tau \qquad\qquad (2\text{-}8)$$

where k/δ = *thermal conductivity* (the inverse of this ratio is the *thermal resistance of the wall*)

   A = heat transfer area

Heat Conduction Through a Cylindrical Wall

This system is illustrated in Figure 2-4. The cylinder has length ℓ, inside radius r_1, and outside radius r_2. The conductivity of the material k is constant. At steady-state the temperatures t_{w1} and t_{w2} $(t_{w1} > t_{w2})$

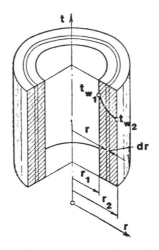

Figure 2-4. Heat conduction through a cylindrical wall.

(i.e., inside and outside surface temperatures) are constant. In this case, the temperature field is one dimensional; temperature changes only in the direction of the radius.

The amount of heat passing through the circular layer inside the wall of thickness dr and radius r may be determined again from Fourier's equation to give:

$$dQ = -k\frac{\partial t}{\partial n}Ad\tau = -k2\pi r\ell\frac{dt}{dr}d\tau \qquad (2\text{-}9)$$

For a steady-state thermal regime, this becomes:

$$Q = -k\pi r\ell\frac{dt}{dr}\tau$$

or

$$dt = -\frac{Q}{2\pi\ell k\tau}\left(\frac{dr}{r}\right) \qquad (2\text{-}10)$$

Integrating over the limits from r_1 to r_2 and from t_{w1} to t_{w2} gives the conductivity equation for a cylindrical wall undergoing steady-state heat conduction:

$$t_{w2} - t_{w1} = -\frac{Q}{2\pi\ell k\tau}2.3\ln\frac{r_2}{r_1}$$

or

$$Q = \frac{2\pi\ell}{\frac{1}{k}2.3\ln\frac{r_2}{r_1}}(t_{w1} - t_{w2})\tau \qquad (2\text{-}11)$$

Heat Conduction Through a Spherical Wall

This system is illustrated in Figure 2-5. This is also a one-dimensional steady-state problem when the interior and exterior surface temperatures are uniform and constant. The sphere has an inside radius r_1, an exterior radius r_2, and conductivity k. The temperature distribution is one dimensional, and the temperature changes only along the direction of the radius.

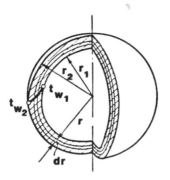

Figure 2-5. Heat conduction through a spherical wall.

According to Fourier's law, the amount of heat passing through a spherical layer of thickness dr and radius r inside the wall is:

$$Q = -k\frac{dt}{dn}A\tau = -k4\pi r^2\frac{dt}{dr}\tau \tag{2-12}$$

or

$$dt = -\frac{Q}{4\pi k\tau}\frac{dr}{r^2} \tag{2-13}$$

Integrating over the limits from r_1 to r_2 and from t_{w1} to t_{w2} gives the conductivity equation for a spherical wall at steady-state heat conduction:

$$t_{w2} - t_{w1} = -\frac{Q}{4\pi k\tau}\left(\frac{1}{r_1} - \frac{1}{r_2}\right)$$

or

$$Q = \frac{4\pi k}{\dfrac{1}{r_1} - \dfrac{1}{r_2}}(t_{w1} - t_{w2})\tau \tag{2-14}$$

Heat Conduction Through Composite Walls

The walls of heat transfer equipment are often constructed of several layers of different substances of varying thicknesses. (see Figure 2-6). The most general one-dimensional case is a flat multilayer wall composed of n layers of thicknesses $\delta_1, \delta_2, \ldots, \delta_n$ and conductivities $k_1, k_2, \ldots, k_n$.

The temperatures on the opposite sides of the multilayer walls are t_{w1} and t_{w2} ($t_{w1} > t_{w2}$), and temperatures on the boundaries between layers $t_1, t_2, \ldots, t_{n-1}$. For steady state the heat flow is the same for all layers. The Fourier equation can be written for each of these layers as follows:

For the first layer:

$$Q = \frac{k_1}{\delta_1} A\tau(t_{w1} - t_1) \text{ or } \frac{Q}{A\tau}\frac{\delta_1}{k_1} = (t_{w1} - t_1)$$

For the second layer:

$$Q = \frac{k_2}{\delta_2} A\tau(t_1 - t_2) \text{ or } \frac{Q}{A\tau}\frac{\delta_2}{k_2} = (t_1 - t_2)$$

For the nth layer:

$$Q = \frac{k_n}{\delta_n} A\tau(t_{n-1} - t_{w2}) \text{ or } \frac{Q}{A\tau}\frac{\delta_n}{k_n} = (t_{n-1} - t_{w2})$$

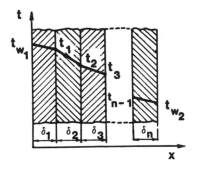

Figure 2-6. Conduction in a multilayer flat wall.

Adding these equations gives the equation of conduction for a multi-layer flat wall:

$$\frac{Q}{A\tau}\left(\frac{\delta_1}{k_1} + \frac{\delta_2}{k_2} + \dots + \frac{\delta_n}{k_n}\right) = (t_{w1} - t_{w2})$$

or

$$Q = \frac{1}{\displaystyle\sum_{i=1}^{n} \frac{\delta_1}{k_i}} A(t_{w1} - t_{w2})\tau \tag{2-15}$$

where i = number of a layer
       n = number of layers

Equation 2-15 states that the overall thermal resistance of a multilayer flat wall is the sum of the individual thermal resistances. For a multilayer cylindrical wall undergoing steady-state heat conduction, the following equation applies:

$$Q = \frac{2\pi\ell}{\displaystyle\sum_{i=1}^{n} \frac{2.3}{k_i} \ln \frac{r_i + 1}{r_i}} (t_{w1} - t_{w2})\tau \tag{2-16}$$

And the thermal conduction through a multilayer spherical wall at steady state is:

$$Q = \frac{4\pi}{\displaystyle\sum_{i=1}^{n} \frac{1}{k_i}\left(\frac{1}{r_i} - \frac{1}{r_i + 1}\right)} (t_{w1} - t_{w2})\tau \tag{2-17}$$

We may now look at two practical heat transfer problems, namely, heat losses from insulated piping, and losses through furnace walls.

Heat Losses from Insulated Piping

Radial heat flow through a cylindrical pipe is basically a one-dimensional, steady-state transfer of energy by conduction. In most cases, axial heat flow can be assumed to be negligible.

In general, flow of heat through a solid pipe wall, can be described by (see Figure 2-7):

$$q = \frac{T_{max} - T}{\Sigma R_T} \qquad (2\text{-}18)$$

where
- q = heat transfer rate, Btu/hr
- T_{max} = maximum temperature in system, °F
- T = temperature of surface toward which heat is flowing, °F
- R_T = thermal resistance of solid per unit thickness, °F/Btu/(hr)(sq ft)

Thermal resistance is analogous to electrical resistance. The reciprocal of electrical resistance is conductance; similarly, the reciprocal of thermal resistance is thermal conductance. Thermal resistance or conductance is a characteristic of the piping material. The summation sign Σ in the denominator of Equation 2-18 indicates that there may be several solids through which heat is being conducted, each with a different thermal resistance.

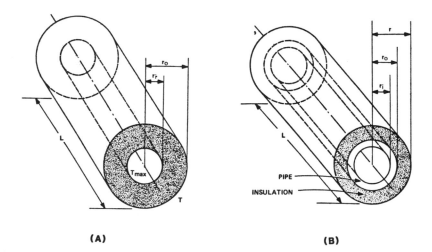

(A) **(B)**

Figure 2-7. (A) Heat losses through single wall pipe; (B) heat losses through insulated pipe.

For cylindrical pipes, thermal resistance R_T is:

$$R_T = \frac{1}{2\pi kL} \ln\left(\frac{r_o}{r_i}\right) \qquad (2\text{-}19)$$

where k = thermal conductivity, Btu/(hr)(sq ft)(°F/ft)
$\quad\quad\quad L$ = length of pipe, ft
$\quad\quad\quad r_o$ = outside pipe radius, ft
$\quad\quad\quad r_i$ = inside pipe radius, ft

Substituting Equation 2-19 for R_T in Equation 2-18 gives:

$$q = \frac{2\pi kL(T_{max} - T)}{\ln\dfrac{r_o}{r_i}} \qquad (2\text{-}20)$$

This equation may, in some cases, be more conveniently expressed in terms of the surface areas of the inner and outer surfaces of the pipe. These areas may be expressed in several ways:

Arithmetic mean area A_A (sq ft):

$$A_A = \frac{A_i + A_o}{2} \qquad (2\text{-}21)$$

where $A_i = 2\pi r_i L$ = inside surface area of pipe, sq ft
$\quad\quad\quad A_o = 2\pi r_o L$ = outer surface area of pipe, sq ft

Geometric mean area A_G (sq ft):

$$A_G = \sqrt{A_i A_o} \qquad (2\text{-}22)$$

Logarithmic mean area A_L (sq ft):

$$A_L = \frac{A_o - A_i}{\ln\left(\dfrac{A_o}{A_i}\right)} \qquad (2\text{-}23)$$

Table 2-2 lists the various ways of expressing the heat transfer rate q in terms of these areas. The first equation in Table 2-2 gives the exact

Table 2-2
Expressions for Heat Transfer Rate q*[4]

Calculation Basis	Equation
Exact value	$q = \dfrac{2\pi LK(T_{max} - T)}{\ln \dfrac{r_o}{r_i}}$
In terms of arithmetic mean area, A_A	$q = \dfrac{2\pi K A_A(T_{max} - T)}{(r_i + r_o)/\ln\left(\dfrac{r_o}{r_i}\right)}$
In terms of geometric mean area, A_G	$q = \dfrac{A_G K(T_{max} - T)}{\sqrt{r_o r_i}\,\ln\left(\dfrac{r_o}{r_i}\right)}$
In terms of logarithmic mean area, A_L	$q = \dfrac{K A_L(T_{max} - T)}{r_o - r_i}$
In terms of the outer or inner surface area (this equation is to be used only for approximate values of q)	$q = \dfrac{KA(T_{max} - T)}{r_o - r_i}$

expression for heat transfer by conduction. The expression in terms of the logarithmic mean area will give approximately the same result.

To express the heat transfer rate when a composite material is involved or when a combination of conduction and convection controls the transfer, use the overall heat-transfer coefficient U. This is expressed as:

$$q_T = UA_s\Delta t = UA_s(T_{max} - T) \tag{2-24}$$

where q_T = total rate of heat transfer, Btu/hr
  U = overall heat-transfer coefficient (sometimes referred to as overall thermal conductance), Btu/(hr)(sq ft)(°F)
  A_s = surface area available to heat transfer, sq ft
  Δt = overall temperature difference between hot and cold fluids, °F

The previous discussion dealt with heat transfer calculations for steady-state conduction. As approximations, these procedures are valid for describing heat losses or gains through the pipe. However, for a more accurate picture of heat transferred through a pipe, convection must be considered.

Setting Equations 2-18 and 2-24 equal and solving for U gives:

$$U = \frac{1}{A_s \Sigma R} \tag{2-25}$$

or more specifically, for a pipe without insulation:

$$U = \frac{1}{A_s \left[\dfrac{1}{A_i h_i} + \dfrac{\ln (r_o/r_i)}{2\pi k_o L} + \dfrac{1}{A_o h_o} \right]} \tag{2-26}$$

where A_s = surface area (most commonly expressed in terms of the inner or outer surface area of the pipe), sq ft

A_i, A_o = inside and outside areas, respectively, of pipe, sq ft

h_i, h_o = convective heat-transfer coefficients of the fluids inside and outside the pipe, respectively, Btu/(hr)(sq ft)(°F)

Equations 2-25 and 2-26 indicate that several factors are important in determining U:

1. Thermal conductivities of the piping wall and insulation.
2. Thickness of the piping wall and insulation.
3. Conductance (convective heat-transfer coefficients) of the fluid and surroundings.

Heat Losses Through Furnace Walls

The heat generated in an industrial furnace is transmitted to the charge, the surrounding walls, and the stack. This subsection considers only the heat transferred by conduction through the wall under steady conditions, no attention being given to the effects of infiltration, furnace pressure, open joints, doors, etc.

Heat is transmitted through refractory walls mainly by conduction, and is then transmitted to the surrounding air and neighboring objects by radiation and convection from the outer wall surface. The drop in temperature through the wall depends upon the resistance offered by the wall to the flow of heat.

The quantity of heat transmitted by conduction through the wall per unit area varies directly with the temperature difference between the outer and inner surfaces of the wall, and the thermal conductivity of the refractory and insulation, and inversely with the thickness. Thermal conductivities of refractory and insulating materials are usually given as a function of mean and average temperature. Figure 2-8 provides an average curve of thermal conductivity for firebrick, common brick, and plastic cement. Table 2-3 gives approximate thermal conductivities of different refractory materials.

The heat that passes through the wall is dissipated to the surrounding air and objects. The rate of heat transfer depends upon various factors, such as the temperature of the surface and the surroundings, the rate of circulation of air over the wall, and the nature of the surface. The total rate of heat transfer through the wall depends upon the thermal resistance of the wall, as well as the resistance to heat dissipated from the surface of the wall. Therefore, before the actual transfer through the wall can be determined, it is necessary to solve for the simultaneous flow of heat through the wall and the surface. The true value of the heat flow is obtained when the solution gives the same heat flow through the wall (by conduction) and from its outer surface (by radiation and convection).

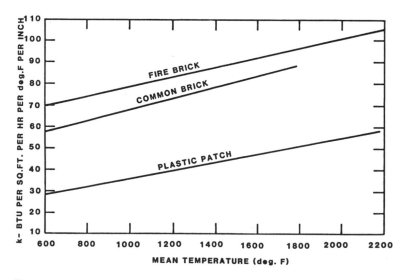

Figure 2-8. Thermal conductivity for firebrick, common brick and plastic patch.

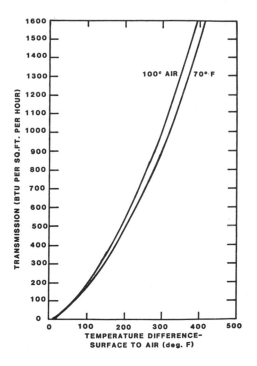

Figure 2-9. Heat transmission from vertical surfaces.

To calculate the heat flow through furnace walls for equilibrium conditions, assume a temperature for the outer wall surface, and from an assumed mean temperature of the various wall components and their conductivities, compute the heat flow through the walls. The calculated heat flow through the wall is then compared with the heat loss from the outer surface due to radiation and convection, with the latter based on the same outside surface temperature. If the two losses are equal and the mean temperature for the wall components check, the calculations are then complete.

The formulas for calculating the radiation or convection from the surface are complex. To facilitate the determination of the surface heat losses, the radiation and convection formulas have been combined and the results of calculations for various temperature differences between the wall surface and surrounding air are shown in Figure 2-9. The plot is based on heat losses from vertical surfaces with surrounding air tem-

Table 2-3
Approximate Thermal Conductivity for Furnace Insulation
(Btu per ft² per hr per °F per in. thickness)

Material	Mean Temperature (°F)													
	100	200	400	600	800	1000	1200	1400	1600	1800	2000	2200	2400	2600
Fireclay brick	5.9	6.2	6.6	7.0	7.5	7.9	8.3	8.8	9.2	9.7	10.1	10.5	11.0	11.4
Common brick	4.6	4.8	5.3	5.8	6.3	6.8	7.4	7.9	8.4					
Magnesite brick	38.1	37.2	35.4	33.3	31.5	30.0	28.3	26.9	25.9	25.0	24.0	23.5	23.2	23.0
Chrome brick	9.3	9.5	10.0	10.5	11.0	11.5	12.0	12.5	13.0	13.5	14.0	14.5	15.0	15.5
Silica brick	6.4	6.8	7.5	8.3	9.0	9.8	10.5	11.3	12.0	12.8	13.5	14.3	15.0	15.8
Silicon carbide						104.0	105.0	106.0	106.5	107.0	108.0	109.0	109.5	110.0
APG plastic firebrick				3.0	3.3	3.6	4.0	4.4	4.8	5.2	5.5	5.9	6.3	
APG KAST-SET or HYDROSET				3.8	3.9	4.0	4.1	4.2	4.2	4.3	4.4	4.5		
APG KAST-O-LITE		1.6	1.7	1.8	1.9	2.0	2.2	2.4	2.6	2.9	3.1			
APG castable insulation No. 20		1.44	1.45	1.48	1.49	1.53	1.57	1.62	1.67	1.73				
APG block insulation		.42	.50	.60	.74	.89	1.08	1.30						
APG insulating cement		.65	.69	.76	.84	.94	1.05	1.19						
APG castable block mix		.53	.56	.60	.67	.75	.85	1.0						
APG loose wool	.28	.36	.54	.77	1.01									
APG G-26 I.F.B.		1.7	1.8	1.9	2.0	2.1	2.2	2.4	2.5	2.7	2.8			
APG G-23 I.F.B.		1.5	1.6	1.7	1.8	1.9	2.0	2.2	2.3	2.4	2.6			

Table 2-3 Continued

Material	Mean Temperature (°F)													
	100	200	400	600	800	1000	1200	1400	1600	1800	2000	2200	2400	2600
Group 28 insulating firebrick		2.38	2.5	2.62	2.75	2.85	3.0	3.13	3.29	3.44	3.60			
Group 20 insulating firebrick		1.16	1.23	1.3	1.4	1.5	1.6	1.73	1.87	2.0				
Natural diatomaceous earth brick right angles to strata		.56	.61	.66	.71	.77	.82	.87						
Natural diatomaceous earth brick parallel to strata		.92	.97	1.02	1.07	1.12	1.17	1.22						
Magnesia 85%	.47	.49	.55											
Asbestos millboard	.85													
Concrete (stone)	6.5													
Sand (dry)	2.3													

Source: A.P. Green Fire Brick Co., Technical Note 3105–B (Dec. 6, 1948).

peratures of 70°F and 100°F., and with an emissivity or radiation constant of 0.95. This radiation constant depends on the nature of the surface, and the value of 0.95 is a good average value for a normal steel furnace casing or brick work. Refer to Sample Calculations 2-4 and 2-15 for illustration.

Notes on Unsteady-State Heat Conduction

Unsteady-state heat conduction is important to many heat transfer applications involving start-ups, operating, and instability transients that may affect the rate at which process equipment reaches stable operation. It is also important in defining the temperature limitations imposed on materials of construction for equipment in terms of heat transfer characteristics, thermal stresses, and temperature distortions. Although it is beyond the present volume to discuss this class of problems, some notes on specific but common problems are worthwhile. The general differential equation for unsteady-state heat transfer by conduction is derived from the basic Fourier equation.

$$\frac{1}{\rho C_P} \left[\frac{\partial}{\partial x}\left(k_x \frac{\partial t}{\partial x}\right) + \frac{\partial}{\partial y}\left(k_y \frac{\partial t}{\partial y}\right) + \frac{\partial}{\partial z}\left(k_z \frac{\partial t}{\partial z}\right) \right] = \frac{\partial t}{\partial \tau} \qquad (2\text{-}27)$$

For many materials, k can be assumed nearly constant over a limited temperature range:

$$\frac{k}{\rho C_P} \left[\left(\frac{\partial^2 t}{\partial x^2}\right) + \left(\frac{\partial^2 t}{\partial y^2}\right) + \left(\frac{\partial^2 t}{\partial z^2}\right) \right] = \frac{\partial t}{\partial \tau} \qquad (2\text{-}28)$$

Integration is possible for only a few cases (one example is an infinite slab).

In an idealized case, at time 0, the slab is at uniform temperature t, and at some later time (when contacted with a coolant), the surface temperature reaches constant t_0. A convenient dimensionless temperature group is:

$$\Delta = \frac{t - t_0}{t_1 - t_0} \qquad (2\text{-}29)$$

Equation 2-29 reflects local temperature and is normalized between the limits of 0 and 1.

Forgoing a rather lengthy derivation, the temperature distribution in an infinite slab is:

$$\Delta = \frac{4}{\pi}\left[\left(e^{-(\pi/2)^2\frac{\alpha\tau}{R^2}}\sin\frac{\pi x}{2R}\right) + \left(\frac{1}{3}e^{-9(\pi/2)^2\frac{\alpha\tau}{R^2}}\sin\frac{3\pi x}{2R}\right)\right.$$

$$\left. + \left(\frac{1}{5}e^{-25(\pi/2)^2\frac{\alpha\tau}{R^2}}\sin\frac{5\pi x}{2R}\right) + \ldots \right] \tag{2-30}$$

Equation 2-30 provides temperature t at any point x, for any time τ for either cooling or heating conditions. The equation is only applicable to evaluating the temperature distribution versus time for a slab heated or cooled on one side and insulated on the other. Note that R denotes the slab thickness.

The amount of heat in the slab is:

$$\frac{Q}{Q_1} = \frac{8}{\pi^2}\left[e^{-(\pi/2)^2\frac{\alpha\tau}{R^2}} + \frac{1}{9}e^{-9(\pi/2)^2\frac{\alpha\tau}{R^2}}\right.$$

$$\left. + \frac{1}{25}e^{-25(\pi/2)^2\frac{\alpha\tau}{R^2}} + \ldots \right] \tag{2-31}$$

where   Q = amount of heat in slab (vs. temperature t_0)
   Q_1 = amount of heat at temperature t_1
   Q_1 = $C(t_1 - t_0)\delta A2R$ \hfill (2-32)
   C = heat capacity
   γ = specific weight
   A = slab surface area

For small process times:

$$\ln\frac{Q}{Q_1} = \ln\frac{8}{\pi^2} - \left(\frac{\pi}{2}\right)^2\frac{\alpha\tau}{R^2} \tag{2-33}$$

Equation 2-33 can be expressed on semi-log coordinates as ln Q vs. τ as a straight line.

Similar expressions have been developed for the transient temperature fields in other geometries (e.g., infinite cylinder or sphere and a body limited by one plane). These equations have been integrated for various boundary conditions and are available in graphical form [1,2]. The dimensionless temperature ratio in these equations is a function of two dimensionless groups:

Fourier modulus (dimensionless time):

$$Fo = \frac{\alpha\tau}{R^2} = \frac{k\tau}{\gamma C R^2} \qquad (2\text{-}34)$$

Dimensionless size:

$$n = \frac{r}{R} \qquad (2\text{-}35)$$

where r is the distance from a given point to the plane of symmetry for a slab as the symmetry axis for a cylinder.

From distance x as measured from one surface and the cylinder radius or half of the slab thickness R we obtain r as an absolute value from R − x. This simply means that the temperature distribution is a function of only two parameters:

$$\Delta = f(Fo, n) \qquad (2\text{-}36)$$

This is valid for simple bodies such as infinite cylinders, infinite plates, and spheres. This is quite useful since many heat transfer surfaces can be approximated by such geometries. For an infinite slab and cylinder, a bundle of lines in the coordinate system Δ- Fo is obtained where each line corresponds to a definite n-value. For a body limited from one side, only one curve is obtained. The body surface is assumed to have temperature t_0 at all times. At very high flows past the surface, heat transfer by convection can be viewed as corresponding to an infinitely great heat transfer coefficient for the slab. An example is when heating metal by saturated steam. In the case of heating (or cooling) with gases of low heat transfer coefficients, the surface temperature of the body will not be equal to that of gas t_0. Even for a high flowrate its temperature t_0 will be constant but the surface temperature of the body will change. If a different heat flow rate passes between the surface and liquid in the unsteady-state regime and if a heat transfer coefficient and one of the temperatures are constant, the other temperature should change, i.e., the surface temperature of the wall.

The thickness of an equivalent layer is equal to the ratio of the material conduction (cylinder, sphere, slab) to the heat transfer coefficient:

$$\delta_z = \frac{k}{h_k} \qquad (2\text{-}37)$$

In some cases there is interest in the temperature distribution only in the part R; i.e., along the slab thickness from a moment when at distance R the temperature is t_1. The ratio of the slab's (cylinder, sphere) cross-section to the thickness of an equivalent layer δ_z is important and is expressed by the dimensionless Biot number.

$$Bi = \frac{\bar{h}R}{k} \qquad (2\text{-}38)$$

Accounting for convection between the slab and heating or cooling flow, Δ is:

$$\Delta = f(Fo,Bi,n) \qquad (2\text{-}39)$$

rather than a function of two variables.

Derivation of this function for an infinite slab, infinite cylinder, and semi-limited body are given by Gurney.[1] Figure 2-10 is presented by

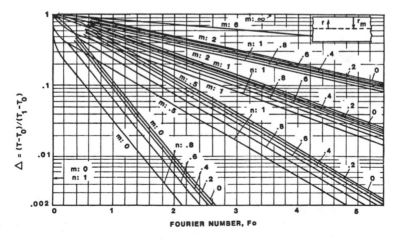

Figure 2-10. Dimensionless groups for a solid unlimited slab.[3]

the function Δ vs. Fo, n, and the inverse of the Biot number m = Bi. There are several families of curves. Each family is determined by the boundary conditions, i.e., whether or not a convective resistance (represented by a heat transfer coefficient h) exists at the surface. Below the families of curves is located a family of curves at m = 0, which relates to ideal conditions of convection. At m = ∞ Bi = 0, i.e., at infinitely slow convection, the temperature of the slab will always be equal to t_1 or Δ = 1 at any value n and Fo (space and time). Each of the curves in the family of curves is related to a definite location in the slab (to a definite n value). At m = 0 (ideal convection) the temperature of the slab surface is equal to t_0, therefore the value Δ at n = 0 is not given here because the corresponding line will coincide with the vertical axis of the system (at Fo = 0).

The case of an infinite cylinder is given in Sample Calculation 2-7. Note that each curve corresponds to a definite value, m or Bi, i.e., to a definite regime of convection. Using these charts we may determine the temperature in the respective geometric bodies depending on the location (n), time (Fo), and convection intensity (m). The case of a sphere is given in Figure 2-11.

Sample Calculation 2-1. (Heat Losses Through a Wall). Estimate the heat losses through a wall having dimensions of 6 m $\times$ 3 m $\times$ 250 mm (L $\times$ H $\times$ W). The surface temperatures are 20°C and −30°C, and the wall's average thermal conductivity is 0.6 kcal/m-hr-°C.

Solution.

$$q = \frac{k}{\delta}(t_1 - t_2)$$

where k = thermal conductivity
       δ = wall thickness

$$q = \frac{0.6}{0.25}[20 - (-30)] = 120 \text{ kcal/m}^2\text{-hr}$$

Sample Calculation 2-2. (Heat Losses Through Composite Wall). Estimate the heat flux through a 20 mm thick wall having a thermal conductivity of 50 kcal/m-hr-°C. The inside surface of the wall is covered with a surfacing material that is 2 mm thick and has an average thermal conductivity of 1.0 kcal/m-hr-°C. The temperatures of the inside and outer surfaces are 200°C and 250°C, respectively.

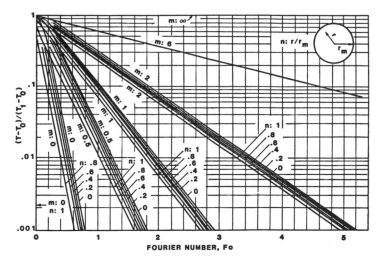

Figure 2-11. Dimensionless groups for a solid sphere.[3]

Solution.

$$q = \frac{t_1 - t_3}{\dfrac{\delta_1}{k_1} + \dfrac{\delta_2}{k_2}} = \frac{250 - 200}{\dfrac{0.02}{50} + \dfrac{0.002}{1}} = 20{,}800 \text{ kcal/m}^2\text{-hr}$$

The temperature of the inside surface (under the substrate) is:

$$t_2 = t_1 - q\frac{\delta_1}{k_1} = 250 - 20{,}800 \times 0.0004 = 241.7\,°C$$

Sample Calculattion 2-3. (Heat Losses Through Pipe Insulation). Estimate the heat losses per m of steam pipe length and the temperatures on the interface of the insulation layers. The steam line has a diameter of 170/160 mm and has two layers of insulation ($\delta_2 = 30$ mm, $\delta_3 = 50$ mm thickness). The thermal conductivities of the pipe and two insulating materials are respectively, $k_1 = 50$, $k_2 = 0.15$, and $k_3 = 0.08$ kcal/m-hr-°C. The inside pipe wall temperature is 300°C (t_1) and that of the outside insulating surface is 60°C (t_4).

Solution. The diameters of the various layers are:

$$d_1 = 0.16 \text{ m}$$
$$d_2 = 0.17 \text{ m}$$
$$d_3 = 0.23 \text{ m}$$
$$d_4 = 0.33 \text{ m}$$

$$q_e = \frac{2\pi(t_1 - t_2)}{\dfrac{1}{k_1}\ln\dfrac{d_2}{d_1} + \ln\dfrac{d_3}{d_2} + \dfrac{1}{k_3}\ln\dfrac{d_4}{d_3}}$$

$$\ln\left(\frac{d_2}{d_1}\right) = 0.06; \quad \ln\left(\frac{d_3}{d_2}\right) = 0.302; \quad \ln\left(\frac{d_4}{d_3}\right) = 0.362$$

$$q_e = \frac{2\pi(300 - 60)}{\dfrac{0.06}{50} + \dfrac{0.302}{0.15} + \dfrac{0.362}{0.08}} = 230.6 \text{ kcal/m-hr}$$

$$t_2 - t_1 - \frac{q_e}{2\pi} \times \frac{1}{k_1}\ln\frac{d_2}{d_1} = 300 - \frac{230.6}{2\pi} \times 0.0012 \simeq 300°C$$

$$t_3 = t_2 - \frac{q_e}{2\pi}\frac{1}{k_2}\ln\frac{d_3}{d_2} = t_1 - \frac{q_e}{2\pi}\left(\frac{1}{k_1}\ln\frac{d_2}{d_1} + \frac{1}{k_2}\ln\frac{d_3}{d_2}\right)$$

$$= 300 - \frac{230.6}{2\pi} \times 2.02 = 300 - 74 = 226°C$$

Sample Calculation 2-4. (Heat Losses Through Furnace Walls). The walls of a furnace consist of three layers: fire-brick (thickness $\delta_1 = 130$ mm), insulating brick ($\delta_2 = 70$ mm) and an outer steel shell ($\delta_3 = 10$ mm). The thermal conductivities of each layer are as follows:

Inner wall: $k_1 = 0.81$ Wt/m-°C (or 0.7 kcal/m-hr-°C)
 Insulating
      brick: $k_2 = 0.23$ Wt/m-°C (0.2 kcal/m-hr-°C)
Steel shell: $k_3 = 45$ Wt/m-°C (40 kcal/m-hr-°C)

The temperature inside the furnace is 810°C and the outside ambient temperature is 30°C. The heat transfer coefficients for the inside and outermost surfaces of the furnace are 69.6 wt/m²-°C (60 kcal/m²-hr-°C) and 13.9 wt/m²-°C (12 kcal/m₂-hr-°C), respectively. Estimate the heat losses per unit area of furnace wall and the overall heat transfer coefficient for a wall.

Solution.

• Compute heat resistances:

For the hot heat carrier:

$$r_1 = \frac{1}{k_1} = \frac{1}{69.6} = 0.0144 \text{ m}^2\text{-°C/Wt}$$

For first layer:

$$r_{w1} = \frac{\delta_1}{k_1} = \frac{0.130}{0.81} = 0.161$$

For second layer:

$$r_{w2} = \frac{\delta_2}{k_2} = \frac{0.070}{0.23} = 0.304$$

For third layer:

$$r_{w3} = \frac{\delta_3}{k_3} = \frac{0.010}{45} = 2.22 \times 10^{-4}$$

For cold heat-carrier:

$$r_2 = \frac{1}{k_2} = \frac{1}{13.9} = 0.072$$

The total thermal resistance is:

$$r_T = 0.0144 + 0.161 + 0.304 + 2.22 \times 10^{-4} + .072$$

$$r_T \simeq 0.552 \text{ m}^2\text{-°C/Wt}$$

- Compute the heat flux equivalent to heat loss per $1m^2$ of wall surface:

$$q = \frac{\Delta t}{r_T} = \frac{810 - 30}{0.552} = 1{,}413 \text{ Wt/m}^2 \text{ (or } 1{,}219 \text{ cal/m}^2\text{-hr)}$$

- Compute the overall heat transfer coefficient:

$$U = \frac{1}{r_T} = \frac{1}{0.552} = 1.81 \text{ Wt/m}^2\text{-°C} \left(\text{or } 1.56 \frac{\text{cal}}{\text{m}^2\text{-hr-°C}} \right)$$

Sample Calculation 2-5. (Estimating Heat Transfer Area and Cooling Fluid Rate). A liquid having a specific heat of 0.8 kcal/kg-°C (3,350 J/kg-°C) is flowing at a mass rate of 12,000 kg/hr. The fluid is to be cooled from 93°C to 38°C. The initial temperature of cooling water is $t_1 = 25$°C. The heat transfer coefficient $h = 290$ wt/m² × °C (250 kcal/m² × hr × °C). Estimate the required heating area and the cooling water rate at cocurrent and countercurrent flow.

Solution.

- First compute the heat load:

$$Q = Q_{hot} = GC(T_1 - T_2) = \frac{12{,}000}{3{,}600} \times 3{,}350 \ (93 - 38)$$

$$= 614{,}167 \text{ Wt}$$

- For cocurrent flow the final temperature of the water t_2 cannot exceed that of the final temperature of the liquid (38°C). Assuming $t_2 = 35$°C, compute the amount of cooling water ($C_p = 1$ kcal/kg-°C or 4,190 J/kg-°C for water):

$$q = \frac{Q}{C(t_2 - t_1)} = \frac{614{,}167}{4{,}190(35 - 25)}$$

$$= 14.7 \text{ kg/s} = 52{,}769 \text{ kg/hr}$$

The average temperature gradient at cocurrent flow is:

$$
\begin{array}{ccc}
93 & \longrightarrow & 38°C \\
25 & \longrightarrow & 35 \\
\hline
\theta_1 = 68 & \theta_2 = & 3
\end{array}
$$

$$\theta_{av} = \frac{68 - 3}{2.3 \ln \frac{68}{3}} = 20.9°C$$

• Compute the heat transfer area for cocurrent flow:

$$A = \frac{Q}{h\theta_{av}} = \frac{614,167}{290 \times 20.9} = 102 \ m^2$$

• Assuming the final temperature of the water is 35°C (the flowrate of water does not change) the average temperature gradient will be:

$$\begin{array}{ccc} 93 & \longrightarrow & 38 \\ 35 & \longleftarrow & 25 \\ \hline \theta_1 = 58 & & \theta_2 = 13 \end{array}$$

$$\theta_{av} = \frac{58 - 13}{2.3 \ln \frac{58}{13}} = 30.1°C$$

• The heat transfer area for countercurrent flow is:

$$A = \frac{614,167}{290 \times 30.1} = 70.4 \ m^2$$

• The above calculations show that for the countercurrent case having the same cooling water rate, the required heat transfer area is less than that of cocurrent flow. The temperature t_2 may be assumed higher than that of the cocurrent flow. This makes it possible to decrease the flow rate of the cooling water. Assuming $t_2 = 50°C$, then the water flow rate and average temperature gradient will be: $G = 21,107$ kg/hr and $\theta_{av} = 25.1°C$ which can be checked by the reader. In this case, the heating area needed is:

$$A = \frac{614,167}{290 \times 25.1} = 84.4 \ m^2$$

Consequently, the cooling water rate may be significantly decreased (by a factor of 2.5) using the countercurrent flow arrangement rather than a cocurrent flow. Hence a smaller heat transfer area is obtained. A further increase in the cooling water temperature t_2 will decrease the rate, however, the heating surface area requirements will increase.

Sample Calculation 2-6. (Heat Loss and Wall Temperature Determination). A ceramic tube (60 mm I.D., 90 mm O.D.) is covered with an outer layer of insulation that is 50 mm thick. A liquid having a mean bulk temperature of 95°C is flowing through it. The outside surface temperature of the insulating layer is $t_2 = 35°C$. The heat transfer coefficient inside the tube is $h_1 = 115$ kcal/m²-hr-°C and the tube's conductivity is $k_1 = 0.9$. The insulating material's conductivity is $k_2 = 0.05$ kcal/m-hr-°C. Compute the heat losses per unit length of tube and the temperature of the inside tube.

Solution.

$$d_c = \frac{0.09 - 0.06}{\ln\left(\frac{.09}{0.06}\right)} = 0.0740 \text{ m}$$

• The equivalent diameter of the insulation is:

$$d_{ins} = \frac{0.19 - 0.09}{\ln\left(\frac{0.19}{0.09}\right)} = 0.1338 \text{ m}$$

where the outside diameter of the insulating layer is $0.09 + 2 \times 0.05 = 0.19$ m.

• The overall heat transfer coefficient is related to the inner tube surface. The heat transfer coefficient for the liquid is $h_1 = 115$ kcal/(m²-hr-°C). For the ceramic wall ($\delta_1 = (90 - 60)/2 = 15$ mm $= 0.015$ m)

$$h_c = \frac{0.9}{0.015} = 60 \frac{\text{kcal}}{\text{m}^2\text{-hr-°C}}$$

The heat transfer coefficient for the insulation layer is:

$$h_{ins} = \frac{0.05}{0.05} = 1 \frac{\text{kcal}}{\text{m}^2\text{-hr-°C}}$$

The overall heat transfer coefficient:

$$\frac{1}{h} = \frac{1}{115} + \frac{0.06}{60 \times .0740} + \frac{0.06}{1 \times .1338} = 0.4706$$

or

$$h = \frac{1}{0.4706} = 2.13 \frac{kcal}{m^2\text{-}hr\text{-}°C}$$

• The temperature gradient between the liquid and the outside surface of the insulation layer is:

$$\Delta t = 95 - 35 = 60°C$$

• The overall coefficient is related to the internal tube surface so that:

$$A = \pi \times 0.06 \times 1 = 0.188 m^2/m$$

Hence, the heat losses per tube unit length of pipe are:

$$q = 2.13 \times 60 \times 0.188 = 24.0 \frac{kcal}{m\text{-}hr}$$

• The outside and inside temperatures of the ceramic tube wall are computed using the following formula:

$$h(t_o - t_n)A_o = h_1(t_o - t_1)A_{av,1} = h_2(t_1 - t_2)A_{av,2}$$

In calculating the outer temperature, the average surface of the insulating layer should be considered based on the equivalent diameter:

$$A_o = \pi \times 0.1338 \times 0.420 m^2/m$$

$$24.0 = 1 \times .420(t_2 - 35)$$

or

$$t_2 = \frac{24.0}{0.420} + 35 = 92.1°C$$

- The temperature of the inside surface of the tube is also computed based on the equivalent surface of this layer:

$$A_1 = \pi \times 0.0740 = 0.233 m^2/m$$

- The heat flux equation for this case is:

$$24.0 = 90 \times 0.233(t_1 - 92.1)$$

Hence:

$$t_1 = \frac{24.0}{90 \times 0.233} + 92.1 = 93.2$$

Sample Calculation 2-7. (Drying Time and Temperature Distribution of a Cylinder). Pine logs at a lumber yard (approximately 30 cm in diameter) are dried using saturated steam ($t_0 = 100°C$). This initial treating stage is completed when the average temperature of a log exceeds 90°C. The thermal conductivity of a dry pine log (in the direction perpendicular to the fiber) is $k_1 = 0.13$ kcal/m-hr-°C. Other properties are the heat capacity $C = 0.38$ kcal/kg-°C, and specific weight $\gamma = 550$ kg/m^3. Logs typically enter this stage of the operation with a mean temperature of $t_1 = 20°C$.

Estimate the time required to bring the logs to an average temperature of 90°C, the temperature distribution inside the log at the end of the process, and the temperature gradient along the axis of the log.

Solution. The thermal diffusivity of the material is:

$$\alpha = \frac{k}{C\gamma} = \frac{0.13}{0.38 \times 550} = 0.00063 \ m^2/hr$$

Assuming that condensate formation at steam heating does not greatly alter the thermal diffusivity of wood, Figure 2-12 can be used to obtain an approximate solution. Since the heat transfer coefficient from saturated steam to the wall is large, the curves for m = 0 may be used. This problem is concerned with the time required for the log tempera-

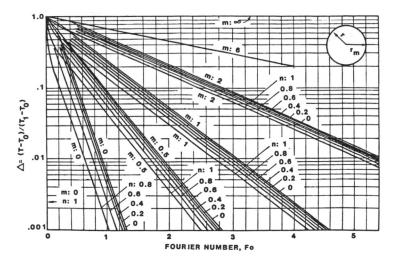

Figure 2-12. Function Δ vs. Fo (Fourier number) and inverse Biot number for an infinite cylinder.

ture to reach 90°C. This temperature will be reached along the log axis at the last moment (r = 0). Consequently, we will use the curve at m = 0 and n = 0.

$$\Delta = \frac{T - T_0}{T_1 - T_0} = \frac{90 - 100}{20 - 100} = 0.125$$

Using the curve for an infinite cylinder at m = 0 and n = 0, we find the corresponding value of Fo (Fo = 0.425). From Fo, determine the time τ. The radius of the log is R = 0.15 m; and the thermal diffusivity of wood is α = 0.00063 m²/hr. Consequently:

$$\tau = \frac{Fo \times R^2}{a} = 0.425 \frac{0.15^2}{0.00063} = 15.2 \text{ hours.}$$

From this time and the Fourier number (Fo = 0.425) Figure 2-12 is consulted for the bundle of curves at m = 0 to obtain values of Δ corresponding to different values of n. After evaluating n, determine the radius r for a given point and knowing Δ we determine the temperature at each point from:

$$t = \Delta(20 - 100) + 100$$

Hence, the values of n and Δ at Fo = 0.425 are obtained.

Using these data, the temperature at any point r from the log's axis can be evaluated after 15 hrs of drying:

n	Δ	r	t
0.0	0.125	0	90
0.2	0.100	3	92.0
0.4	0.085	6	93.2
0.6	0.060	9	95.2
0.8	0.030	12	97.0
1.0	0.000	15	100

It is also possible to determine the temperature change along the log's axis versus time.

For various times, $\tau = 1, 2, 3, \ldots, 15$ hours, the following expression can be used.

$$Fo = \frac{a\tau}{R^2} = \frac{k\tau}{\gamma CR^2}$$

From Figure 2-12, at m = 0 and n = 0, determine corresponding values of Δ. The temperature and Fourier number are computed from:

$$Fo = \frac{0.00063}{0.15^2} \tau$$

The results of these calculations follow:

τ	Fo	Δ	t
2	0.056	0.970	22
4	0.110	0.845	36
6	0.170	0.600	52
8	0.225	0.435	65
10	0.280	0.315	70
12	0.330	0.240	81
14	0.390	0.170	86
15.2	0.425	0.125	90

Sample Calculation 2-8. (Estimating Heating Time). During the manufacture of canned stewed tomatoes, the material is steam heated (t = 115°C) from an initial temperature of the can (t = 15°C) up to t = 105°C. The can diameter is d = 20 cm. The resistance of the can metal is negligible.

The actual conductivity of the canned tomatoes is k = 1.6 kcal/m-hr-°C, and its specific gravity is γ = 1,100 kg/m³; the heat capacity is C = 1.05 kcal/kg-°C.

Compute the heating time.

Solution. Since steam heating is employed and the thermal resistance of the container's wall can be neglected, we may assume that the surface temperature of tomatoes in the can during the entire process is t_0 = 115°C. The can can be considered a combination of an infinite slab and an infinite cylinder. Consequently, the equation for the center of symmetry of the can is:

$$\Delta = \Delta(W)\Delta(P)$$

The ratio of temperature differences is:

$$\Delta = \frac{T - T_o}{T_1 - T_o} = \frac{105 - 115}{15 - 115} = 0.10$$

where T = 105°C is the temperature that should be achieved in the center of the can at the end of the process, and T_1 = 15°C is the initial temperature of the tomatoes.

The thermal diffusivity of the tomatoes is:

$$\alpha = \frac{k}{C\gamma} = \frac{1.60}{1.05 \times 1,100} = 0.00138$$

The temperature change along the axis of an infinite cylinder can be computed from:

$$\Delta = j_\ell^{-2.303(\alpha\tau/R^2t)}$$

values of j and t for different geometries are given in Table 2-4.

The following equation also applies:

$$\Delta(W) = 1.602j_\ell^{-2.303(\alpha\tau/0.398R^2)}$$

Table 2-4
Values of j and t for Different Geometries

Geometry	j	t
Infinite slab	1.273	0.933
Infinite cylinder	1.602	0.398
Sphere	2.000	0.233

where R is the cylinder radius equal to half of the can height (as given).

The change in temperature in the symmetry plane of an infinite slab may be approximated by the equation:

$$\Delta(P) = 1.27 j_\ell^{-2.303(\alpha\tau/0.933R^2)}$$

The temperature at the center of symmetry is:

$$\Delta = \Delta(W)\Delta(P) = 1.273 \times 1.602^{-2.303\frac{\alpha}{R^2}\left(\frac{1}{0.933} + \frac{1}{0.398}\right)}$$

Substituting $\Delta = 0.10$ we obtain:

$$\frac{\alpha\tau}{R^2} = 0.366$$

The thermal diffusivity is $\alpha = 0.00138$, and the can radius R = 0.10 m. Hence, the time required for the process is:

$$\tau = \frac{0.366 \times 0.10^2}{0.00138} = 2.6 \text{ hours}$$

Sample Calculation 2-9. (Drying Problem). The temperature of an apple pie removed from the oven decreases from 100° to 60°C over a period of 25 minutes (refer to Figure 2-13). The ambient air temperature is 25°C. Estimate the time required for the pie's temperature to reach 30°C.

Solution. From Newton's law, the cooling rate is proportional to the temperature difference. This system is an unsteady-state process; that is, with a change of temperature difference during the process, the rate

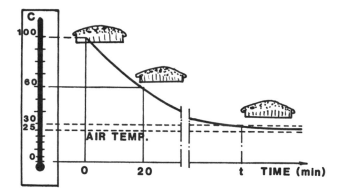

Figure 2-13. System described in Sample Calculation 2-9.

of cooling changes as well. The differential equation of the pie's cooling is:

$$\frac{dT}{d\tau} = K(T - t)$$

where
$$
\begin{aligned}
T &= \text{pie temperature} \\
t &= \text{temperature of ambient air (t = 25°C)} \\
K &= \text{conductivity coefficient} \\
dT/d\tau &= \text{cooling rate of the pie}
\end{aligned}
$$

Denoting τ as the cooling time and dividing the variables, we obtain:

$$\frac{dT}{T - t} = K d\tau$$

or

$$\frac{dT}{T - 25} = K d\tau$$

Upon integration:

$$\int \frac{dT}{T - 25} = K \int d\tau$$

or

$$\ln (T - 25) = K\tau + \ln C$$

Taking antilogarithms of both sides of the last equation we obtain:

$$e^{\ln (T - 25)} = e^{K\tau + \ln C} = e^{K\tau} \times e^{\ln C}$$

Since

$$e^{\ln N} = N$$

then

$$T - 25 = Ce^{K\tau} \qquad (2\text{-}40)$$

The constant C is determined from the initial condition: at $\tau = 0$ min, $T = 100°C$.
Hence:

$$100 - 25 = Ce^{K \times 0} = C \text{ or } C = 75$$

The value e^K is determined from the following boundary condition: at $\tau = 25$ min, $T = 60°C$

$$60 - 25 = 75(e^K)^{25}$$

and

$$e^K = \left(\frac{35}{75}\right)^{1/25} = \left(\frac{7}{15}\right)^{1/25}$$

Thus, the cooling of the pie can be expressed as:

$$T = 75\left(\frac{7}{15}\right)^{\frac{\tau}{25}} + 25 \qquad (2\text{-}41)$$

From Equation 2-41, τ is obtained for $T = 30°C$

$$5 = 75\left(\frac{7}{15}\right)^{\tau/25}$$

or

$$\frac{1}{15} = \left(\frac{7}{15}\right)^{\tau/25}$$

Finally,

$$\tau = \frac{-25 \ln 15}{\ln 7 - \ln 15} \simeq \frac{25 \times 2.7081}{-0.7622} \simeq 89 \text{ min}$$

Thus, after 1 hr and 29 min the pie cools down to 30°C and we can begin eating.

Sample Calculation 2-10. (Temperature Distribution in a Sphere). A 10 mm steel sphere is at a steady thermal state. A miniature thermocouple imbedded in the sphere at a radius of 6 mm senses the temperature to be 200°C. The outer skin temperature is measured as 20°C (see Figure 2-14). Compute the temperature at a distance r from the center and the amount of heat the sphere loses to the outside. The sphere's conductivity is k = 0.14 kcal/m-hr-°C.

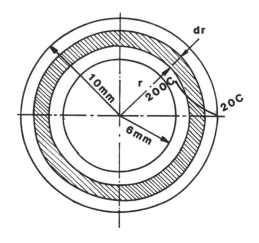

Figure 2-14. System described in Sample Calculation 2-10.

Solution. From symmetry considerations, we may assume that the heat radiates outward. At distance r from the center, the heat transfer area is:

$$A = 4\pi r^2$$

Between two spherical surfaces, the heat rate is constant, and equal to:

$$Q = kA \frac{dT}{dr} \tag{2-42}$$

where T = body temperature
       k = conductivity

Equation 2-42 may be restated as:

$$-4\pi kr^2 \frac{dT}{dr} = Q = \text{const.}$$

Integrating Equation 2-42, we obtain:

$$4\pi kT = \frac{Q}{r} + C \tag{2-43}$$

For a particular solution, substitute the initial conditions T = 20°C, r = 10 mm; T = 200°C, r = 6 mm into Equation 2-43 and determine C and Q:

$$800\pi k = \frac{Q}{1.0} + C$$

$$800\pi k = \frac{Q}{0.6} + C$$

or solving simultaneously:

$$C = -1,000\pi k$$

$$Q = 1,080\pi k$$

The desired temperature at distance r is:

$$T = \frac{1,080}{4r} - \frac{1,000}{4}$$

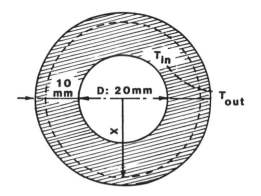

Figure 2-15. System described in Sample Calculation 2-11.

$$T = \frac{270}{r} - 250$$

and the heat flow is:

$$Q = 1,080\pi k = 1,080\pi(0.14) = 475 \text{ kcal/s}$$

Sample Calculation 2-11. (Temperature Distribution and Heat Losses Through Pipe Insulation). A 20 mm I.D. tubing has a layer of insulation 10 mm thick wrapped around it. The insulation's conductivity is $k = 1.9 \times 10^{-4}$ kcal/hr-m-°C. The temperature of the tubing is t = 160°C and that of the external surface of the insulation is 30°C (see Figure 2-15). Compute the temperature distribution inside the insulating layer and the heat losses per unit length of tubing.

Solution. At steady state, the temperature at each point through the tube is a function of only the x coordinate. From Fourier's law, the heat flux is:

$$Q = -kA(x)\frac{dT}{dx} = \text{const.} \tag{2-44}$$

where $A(x)$ = cross-sectional area at a distance x from the centerline
  k = conductivity

$$A(x) = 2\pi x\ell \tag{2-45}$$

where ℓ is the tube length.

The differential form of the equation is:

$$dT = -\frac{Q}{kA(x)} dx = -\frac{Q}{k2\pi\ell} \frac{dx}{x} \qquad (2\text{-}46)$$

By integration both sides of Equation 2-46 we obtain:

$$\int_{160}^{30} dT = -\frac{Q}{0.00017 \times 2\pi\ell} \int_{1}^{2} \frac{dx}{x} \qquad (2\text{-}47)$$

$$\int_{160}^{t} dT = -\frac{Q}{0.00017 \times 2\pi\ell} \int_{10}^{x} \frac{dx}{x} \qquad (2\text{-}48)$$

or

$$T - 160 = -\frac{Q}{0.00017 \times 2\pi\ell} \ln (0.1x) \qquad (2\text{-}49)$$

Dividing the terms in Equation 2-48 by those in Equation 2-47 gives:

$$\frac{T - 160}{-130} = \frac{\ln 0.1}{\ln 2} = \frac{\ln 0.1x}{\ln 2}$$

Hence, the temperature distribution through the insulating layer is:

$$T = 591.8 - 431.8 \ln x$$

From Equation 2-47 at $\ell = 100$ cm, we obtain:

$$Q = \frac{130 \times 0.00017 \times 2\pi \times 100}{\ln 2} = \frac{200\pi \times 130 \times 0.00017}{0.6931}$$

$$= 20.0 \frac{kcal}{s}$$

Sample Calculation 2-12. (Steady-State Conduction Through a Pipe Wall). A standard 2-in. copper pipe with an inside radius of 1.03 in. has hot water at a temperature of 140°F flowing through it. The temperature at the outer pipe surface is 100°F. Determine heat loss q by conduction per inch of pipe length. For copper pipe between 100° and 212°F, the thermal conductivity k is approximately 219 Btu/(hr)(sq ft)(°F/ft). The outer radius of the pipe r_o is 1.18 in.

Solution. Use the expression from Table 2-2 for the exact value of q with L = 1¹/₂ ft. Then:

$$q = \frac{4,587}{\ln (1.146)} = 33,650 \text{ Btu/hr per in. of pipe length}$$

Expressed in terms of the logarithmic mean area:

$$A_L = 0.04803 \text{ sq ft of } 6.916 \text{ sq in.}$$

$$q = 33,660 \text{ Btu/hr per in. of pipe length}$$

For the approximate value (basing calculations on the outside surface area A_o of the pipe):

$$A_o = 2\pi L r_o = 0.05149 \text{ sq ft or } 7.415 \text{ sq in.}$$

$$q = 36,080 \text{ Btu/hr per in. of pipe length}$$

Note that the values for the heat transfer rate obtained from the approximate formula become less accurate as the ratio of r_o/r_i increases (see Table 2-5).

Insulation changes the heat transfer rate by adding thermal resistance. The total thermal resistance of the system is the sum of the resistances of the insulating material and pipe. Figure 2-7B shows a cylindrical pipe with a single layer of insulation. The radial heat loss through this system may be calculated from:

$$q = \frac{2L(T_{max} - T)}{P + I} \qquad (2\text{-}50)$$

where

$$P = \frac{1}{k_p} \ln \left(\frac{r_o}{r_i}\right)$$

$$I = \frac{1}{k_i} \ln \left(\frac{r}{r_o}\right)$$

k_p = thermal conductivity of piping material, Btu/(hr)(sq ft)(°F/ft)

Table 2-5
Effect of Radius Ratio (r_o/r_i) on Calculation Error

Equation Number	% Error
2-19	4
2-20	10
2-21	15

$$q = \frac{kA(T_{max} - T)}{r_o - r_i}$$

k_i = thermal conductivity of insulating material, Btu/(hr)(sq ft)(°F/ft)

r = outside radius of insulation, ft

In general, for any number of layers of insulation:

$$q = \frac{2\pi L(T_{max} - T)}{P + I_1 + I_2 + I_3 \ldots} \qquad (2\text{-}51)$$

where

$$I_1 = \frac{1}{k_{i1}} \ln\left(\frac{r}{r_o}\right)$$

$$I_2 = \frac{1}{k_{i2}} \ln\left(\frac{r'}{r}\right)$$

$$I_3 = \frac{1}{k_{i3}} \ln\left(\frac{r''}{r'}\right)$$

k_{i1}, k_{i2}, k_{i3} = thermal conductivities of successive layers of insulating material, Btu/(hr)(sq ft)(°F/ft)

r, r', r'' = outside radii of successive layers of insulating material, ft

Equations 2-50 and 2-51 both give the exact value for the heat transfer rate. For approximate values, the following general expression may be used:

$$q = \frac{\pi L(T_{max} - T)}{P_a + I_{1a} + I_{2a} + I_{3a} \ldots} \qquad (2\text{-}52)$$

where

$$P_a = \frac{1}{k_p} \left(\frac{r_o - r_i}{r_o + r_i} \right)$$

$$I_{1a} = \frac{1}{k_{i1}} \left(\frac{r - r_o}{r + r_o} \right)$$

$$I_{2a} = \frac{1}{k_{i2}} \left(\frac{r' - r}{r' + r} \right)$$

$$I_{3a} = \frac{1}{k_{i3}} \left(\frac{r'' - r'}{r'' + r'} \right)$$

Sample Calculation 2-13. (Heat Loss Through Insulated Pipe). A steel pipe with an outside radius of $r_o = 2.0$ in. and inside radius of $r_i = 1.03$ in. has steam at a temperature of $806°F$ flowing through it. Three layers of insulation are to be wrapped around the pipe. A 1.5-in.-thick layer of diatomaceous earth is the first layer of insulation around the outer surface of the pipe, a 2.0-in.-thick layer of 85% magnesia cloth is the middle layer, and a 2.5-in.-thick layer of diatomaceous base cement is the outermost layer.

Estimate the heat flow per unit length of pipe if the temperature at the outermost surface of the insulation is $104°F$.

Thermal conductivities, thicknesses, and radii of the insulating materials are tabulated in Table 2-6. The thermal conductivity of steel pipe is 1.04 Btu/(hr)(sq ft)(°F/ft) at 800°F.

Table 2-6
Data for Sample Calculation 2-13

Insulator	Thickness (in.)	Thermal Conductivity k, Btu/(hr)(sq ft) (°F/ft)	Outside Radius (in.)
Diatomaceous earth	1.5	0.0417	$r = 3.5$
85% magnesia cloth	2.0	0.0459	$r' = 5.5$
Cement (diatomaceous base)	2.5	0.0542	$r'' = 8.0$

Solution. The basic expression for these calculations is Equation 2-51. From it, an expression can be written for calculating the exact value of heat loss of a pipe with three layers of insulation:

$$q = \frac{2\pi L(T_{max} - T)}{P + I_1 + I_2 + I_3} \qquad (2\text{-}53)$$

where P, I_1, I_2, I_3, are as defined before. Also,

where $k_p = 1.04$, and from tabulated values above, $k_{i1} = 0.0417$, $k_{i2} = 0.0459$, and $k_{i3} = 0.0542$.

Substituting values in Equation 2-53 and solving for q/L (heat loss per unit length of pipe) gives:

$$P = 0.6382$$
$$I_1 = 13.42$$
$$I_2 = 9.841$$
$$I_3 = 6.919$$

$$\frac{q}{L} = \frac{2\pi(T_{max} - T)}{P + I_1 + I_2 + I_3} = 143.1 \text{ Btu/hr per ft of pipe}$$

Similarly, an expression for calculating the approximate heat loss may be written and solved as follows:

$$q = \frac{\pi L(T_{max} - T)}{P_a + I_{1a} + I_{2a} + I_{3a}} \qquad (2\text{-}54)$$

where $P_a = 0.3078$
$\quad I_{1a} = 6.540$
$\quad I_{2a} = 4.841$
$\quad I_{3a} = 3.417$

$$\frac{q}{L} = 146.0 \text{ Btu/hr per ft of pipe}$$

Sample Calculation 2-14. (Overall Heat Loss through Piping). A stainless-steel pipe with an outside radius (r_o) of 2.0 in. and inside radius (r_i) of 1.03 in. has water at a temperature of 180°F flowing through it. The surrounding air does not circulate. Glass cloth insulation (in the form of a blanket) is to be used as an insulator at a thickness of 1.25 in. The outer surface temperature is approximately 50°F. Determine the overall heat transfer rate to the surroundings.

Values of the heat-transfer coefficients needed for the calculations are: convective heat transfer coefficient of air (stagnant) h_o = 5 Btu/(hr)(sq ft)(°F); convective heat transfer coefficient of hot water (forced convection) h_i = 150 Btu/(hr)(sq ft)(°F); thermal conductivity of piping material k_p = 10.0 Btu/(hr)(sq ft)(°F/ft); and thermal conductivity of insulating material k_i = 0.0260 Btu/(hr)(sq ft)(°F/ft).

Solution. An expression for U for the specific conditions given is:

$$U = \frac{1}{A_s(R_{TW} + R_{TP} + R_{TI} + R_{TA})} \tag{2-55}$$

where $R_{TW} = \dfrac{1}{A_i h_i}$

$$R_{TP} = \frac{\ln\left(\dfrac{r_o}{r_i}\right)}{2\pi k_p L}$$

$$R_{TI} = \frac{\ln\left(\dfrac{r}{r_o}\right)}{2\pi k_i L}$$

$$R_{TA} = \frac{1}{A_o h_o}$$

Note that U may have a different numerical value, depending on whether it is based on the inside or the outside surface area.

If the calculation is based on the inside surface ($A_8 = A_i$), then

$$U = \frac{1}{A_i(R_{TW} + R_{TP} + R_{TI} + R_{TA})} \tag{2-56}$$

Assuming length of pipe L = 1 ft and

$$r = 2.0 + 1.25 = 3.25 \text{ in.}$$
$$A_o = 2\pi r L = 1.702 \text{ sq ft}$$
$$A_i = 2\pi r_i L = 0.5394 \text{ sq ft}$$

$$R_{TW} = 0.01236$$
$$R_{TP} = 0.01056$$
$$R_{TI} = 2.972$$
$$R_{TA} = 0.1175$$
$$U = 0.5956 \text{ Btu/(hr)(sq ft)}(°F)$$
$$q = UA_i (T_{max} - T)$$
$$41.76 \text{ Btu/hr per ft of pipe}$$

Based on outermost surface area $(A_8 = A_o)$,

$$U = 0.1890 \text{ Btu/(hr)(sq ft)}(°F)$$

The heat transfer rate q will have the same value of 41.76 Btu/hr per ft of pipe.

Sample Calculation 2-15. (Heat Loss Through a Two-Component Furnace Wall). A furnace wall consists of 13.5 in. of fire clay brick and 4.5 in. of a commercial insulating firebrick. The inside wall temperature is 2800°F, and the ambient air temperature is 70°F. Estimate the heat loss.

Solution. The formula for heat flow through composite walls is:

$$Q = \frac{t_1 - t_2}{\dfrac{L_1}{k_1} + \dfrac{L_2}{k_2} + \ldots + \dfrac{L_n}{k_n}}$$

where

Q = the quantity of heat conducted in Btu per sq ft per hr

t_1 = temperature of the hotter surface, °F

t_2 = temperature of the cooler surface, °F

$L_1, L_2, \ldots, L_n$ = thickness in inches, of each material

$k_1, k_2, \ldots, k_n$ = conductivity of each material in Btu per sq ft per hr per °F per in. thickness.

To compute the heat loss, assume an outer surface temperature t_2 and estimate the temperature at the junction of each of the wall components from the mean temperature of each material. The thermal conductivities corresponding to the mean temperature of each material should be used in the calculations.

The first trial calculation is generally very approximate. For insulated high-temperature furnaces, the outer surface temperature may be assumed at somewhere between 250° and 350°F. Assume 350°F.

In making the first assumption for the temperatures at the junctions of the materials, guess at these temperatures rather than estimating them by proportion from the relative thickness and relative conductivity of the materials. In this example, assume a temperature of 2000°F between fire clay brick and the commercial material. The corresponding mean temperature of the fire clay brick would be (2,800 + 2,000)/2 or 2400°F, and the corresponding mean temperature of the insulating fire brick would be (2,000 + 350)/2 or 1175°F. The conductivity of the fire clay brick and the commercial brick at these assumed mean temperatures is found on their respective conductivity charts, and a trial calculation made as follows:

$$Q = \frac{2,800 - 350}{\dfrac{13.5}{11.00} + \dfrac{4.5}{2.00}} = \frac{2,400}{1.23 + 2.25}$$

$$= 704 \text{ Btu per sq ft per hr}$$

From the surface loss chart (Figure 2-9), it is found that for a heat loss of 704 Btu, the temperature of the outside surface of the wall will be 243°F above the air temperature. The surface temperature is, therefore, 243 + 70 = 313°F. Since a surface temperature 350°F was first assumed, it will now be desirable to assume a temperature closer to 313°F, say 320°F, and recalculate on that basis. Note that the error in the original estimate of surface temperature not only affected the estimate of interface temperature, but it also influenced the mean temperatures and therefore the thermal conductivity values. This recalculation will reduce errors.

With these recalculations, the temperature drop through the wall will now be 2480°F (that is, 2,800–320) and the heat flow for the new assumed surface temperature will be 2480°F divided by the wall resistance {13.5/11.00 + 4.5/2.00} as just obtained—or 2,480/3.48 = 713 Btu. Since the temperature drop through each material is equal to the heat flow times the actual resistance of each material we obtain the values as tabulated in Table 2-7.

The conductivities of the fire clay and fire brick are again obtained from the manufacturer's conductivity charts for the new mean temperatures, and the heat flow equation then becomes:

$$Q = \frac{2,800 - 320}{\dfrac{13.5}{10.90} + \dfrac{4.5}{1.97}} = \frac{2,480}{1.24 + 2.28} = \frac{2,480}{3.52}$$

$$= 704 \text{ Btu per sq ft per hr}$$

Table 2-7
Tabulated Values for Sample Calculation 2-15

Temperature Drop Through	Hot Face Temperature (°F)	Cold Face Temperature (°F)	Mean Temperature (°F)
Fire Clay Brick $713 \times 1.28 =$ 877 . . .	2800	1923	2361
Common Brick $713 \times 2.25 =$ 1603 . . .	1923	320	1122

From the surface loss chart (Figure 2-9) for a heat loss of 705 Btu, the temperature drop is 244°F, which reflects a surface temperature of $243 + 70 = 313$°F. This is close for all practical purposes to the surface temperature of 320°F assumed previously, therefore, the final answer is a heat loss of 705 Btu per sq ft per hr., and an outside wall temperature of 313°F.

References

1. McAdams, W. H., *Heat Transmission,* McGraw-Hill Book Co., NY (1942).
2. Perry, J. H., *Chemical Engineers Handbook,* 5th Edition, McGraw-Hill Book Co., NY (1963).
3. Gurney, H. P., and J. Lurie, *Ind. Eng. Chem.,* 15:1,170 (1923).
4. Cheremisinoff, N. P., and P. N. Cheremisinoff, *Plant Engineering* (April 29, 1976), pp. 259–263.

Suggested Readings

Barrat, T., *Int. Crit. Tables,* 5:227 (1929).
Bennet, C. O., and J. E. Myers, *Momentum, Heat, and Mass Transfer,* McGraw-Hill, NY (1982).
Brown, G. G., et al., *Unit Operation,* John Wiley, NY (1950).
Buddenberg, J. W., *Ind. Eng. Chem.,* 41 (7):1,345 (1949).
Chapman, A. J., *Heat Transfer,* McMillan Company, NY (1967).
Comings, E. W., *Ind. Eng. Chem.,* 39:964 (1947).

Eckert, E. R. G., and R. M. Drake, Jr., *Analysis of Heat and Mass Transfer,* McGraw-Hill, NY (1972).

Gamson, B. W., *Chem. Eng. Progr.,* 45:154 (1949).

Gurney, H. P., *Ind. Eng. Chem.,* 15:1,170 (1923).

Heisler, M. P., *Trans. ASME,* 69:227 (1947).

Hirschfelder, J. O., *Proc. Joint Conf. Thermodynamic and Transport Properties of Fluids,* The Inst. Mech. Eng. (London), (1958), p. 133.

Holman, J. P., *Heat Transfer,* McGraw Hill, NY (1976).

Jahnke, E., and F. Emde, *Tables of Functions, 4th Ed.,* Dover Publications, NY (1945).

Jakob, M., *Heat Transmission,* McGraw-Hill, NY (1952).

Mesill, D. G., *Ind. Eng. Chem.,* 40:2,263 (1948).

Neuman, A. B., *Trans. Am. Inst. Chem. Eng.,* 27:203 (1931).

Olson, F. S., et al., *Ind. Eng. Chem.,* 34 (336):874 (1942).

Schneider, P. J., *Conduction Heat Transfer,* Addison-Wesley Pub. Co., Reading, MA (1955).

Stolyarov, E. A., et al., *Zh.F.Kh.,* 24:166,279 (1950).

3

CONVECTIVE
HEAT TRANSFER

General Notes

The convective heat transfer coefficient can be stated as follows:

$$dQ = \bar{h}dA(t_w - t_f)d\tau \tag{3-1}$$

Equation 3-1 states that the heat transfer rate dQ per unit time $d\tau$ is directly proportional to the temperature difference between the wall t_w and fluid t_f and to the available heat transfer area dA. For a continuous process over the entire equipment surface:

$$Q = \bar{h}A(t_w - t_f) \tag{3-2}$$

where $\bar{h}$ = convective heat transfer coefficient; units are J/m^2-s-°C or Wt/m^2-°C or $kcal/m^2$-hr-°C. Equation 3-1 is also known as Newton's formula.

The value of $\bar{h}$ depends on many factors; principal ones are fluid velocity u, density ρ, viscosity μ, specific heat C_p, conductivity k, coefficient of thermal expansion β, system diameter d, length L, and wall roughness ϵ.

The most general differential equation of convective heat transfer that expresses the temperature distribution in a flowing fluid is the Fourier-Kirchhoff equation:

$$\frac{\partial t}{\partial \tau} + \frac{\partial t}{\partial x}u_x + \frac{\partial t}{\partial y}u_y + \frac{\partial t}{\partial z}u_z = \alpha \nabla^2 t \tag{3-3}$$

where $\alpha = \dfrac{k}{C\rho}$, thermal diffusivity.

**BOUNDARY
LAYER**

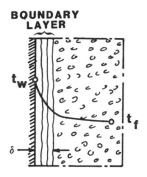

Figure 3-1. Illustration of the temperature distribution for fluid flowing past a vertical surface.

For steady-state processes, $\partial t / \partial \tau = 0$; and for solid bodies, $W_x = W_y = W_z$ from which Equation 3-3 reduces to the differential equation of conduction.

Figure 3-1 illustrates flow past a surface. The temperature distribution in the fluid during convective heat transfer is such that the greatest gradients exist in the boundary layer; the thermal resistance in this region is established mainly by the intensity of heat transfer.

In practical applications, the convective heat transfer coefficient $\bar{h}$ is best expressed in terms of dimensionless groups.

Dimensionless Groups

Nusselt Number

$$Nu = \frac{\bar{h}\ell}{k} \tag{3-4}$$

Equality of Nusselt numbers characterizes the similarity of heat transfer at the interface between wall and fluid in different systems. Nu is basically a measure of the ratio of the boundary layer thickness δ and the system geometric dimension ℓ (e.g., vessel diameter).

Fourier Number

$$\text{Fo} = \frac{\alpha\tau}{\ell^2} \tag{3-5}$$

Equality of Fourier numbers at compatible points of thermal flows is a necessary condition of similarity in unsteady-state heat transfer processes.

Peclet Number

$$\text{Pe} = \frac{u\ell}{\alpha} \tag{3-6}$$

The Peclet number reflects the relative amounts of heat transferred by convection and conduction during convective heat transfer. It can be expressed as the product of two dimensionless groups.

$$\text{Pe} = \frac{u\ell}{\nu} \times \frac{\nu}{\alpha} = \frac{u\ell\rho}{\mu} \times \frac{\mu C_p}{k} = \text{RePr} \tag{3-7}$$

where Re = Reynolds number
       Pr = Prandtl number

Prandtl Number

$$\text{Pr} = \frac{\nu}{\alpha} = \frac{\mu C_p}{k} \tag{3-8}$$

The Prandtl number contains information on the physical properties of the fluid and characterizes the similarity of physical properties of heat carriers in the convective heat transfer process. For liquids, Pr varies between $3 \backsim 300$; for gases Pr typically varies between $0.7 \backsim 1.0$.

Table 3-1 summarizes important dimensionless groups, and Table 3-2 lists common symbols and units used.

Reynolds Analogy

Both heat transfer and fluid friction in turbulent flow can be related through this analogy. The drag on a unit area of pipe wall is:

$$\mu = \tau' = \frac{f}{2}\rho u^2 \tag{3-9}$$

Table 3-1
Important Dimensionless Groups

Symbol	Name	Group
Bi	Biot Number	$h\ell/k$
Fo	Fourier Number	$k\theta/\rho C\ell^2$
Gz	Graetz Number	wC/kL
Gr	Grashof Number	$D^3\rho^2 g\beta\Delta t/\mu^2$
Nu	Nusselt Number	hD/k
Pe	Peclet Number	DGC/k
Pr	Prandtl Number	$C\mu/k$
Re	Reynolds Number	$DG/\mu,\ Du\rho/\mu$
Sc	Schmidt Number	$\mu/\rho k_d$
St	Stanton Number	hC/G

where τ' = drag force
$\quad\ \mu$ = denotes loss of momentum of fluid

The rate of heat transfer between fluid and wall is:

$$mC2\pi rdL(T - t_p) = WCdT \tag{3-10}$$

and

$$m = \frac{WdT}{2\pi rdL(T - t_p)} = \frac{\tau'}{u} \tag{3-11}$$

where W is mass flow
Equation 3-11 states:

$$\frac{\text{heat given up to wall}}{\text{total heat available}} =$$

$$\frac{\text{momentum lost by skin friction to wall}}{\text{total momentum available}}$$

$$= \text{constant}$$

This can be stated as:

$$WCdT = h_i2\pi rdL(T - t_p) \tag{3-12}$$

Table 3-2
Symbols and Dimensions

Symbol	Definition and Typical Units	Dimensions
g	Acceleration of gravity, ft/hr^2, cm/sec^2	L/θ^2
A	Area or surface, ft^2, cm^2	L^2
K_H	Conversion from kinetic energy to heat	$ML^2/H\theta^2$
K_M	Conversion from force to mass	$ML/F\theta^2$
ρ	Density, lb/ft^3, g/cm^3	M/L^3
D	Diameter, ft, cm	L
k_d	Diffusivity (volumetric), ft^2/hr, cm^2/s	L^2/θ
F	Force, force-pound (poundal), force-gram (dyne)	F
H	Heat, Btu, cal	H
L	Length, ft, cm	L
M	Mass,lb,g	M
W	Mass flow, lb/hr, g/s	M/θ
G	Mass velocity, lb/(hr)(ft^2), g/(s)(cm^2)	$M/\theta L^2$
J	Mechanical equivalent of heat, (force-lb)(ft)/Btu, (force-g)(cm)/cal	FL/H
P	Pressure, force-lb/ft^2, force-g/cm^2	F/L^2
P_0	Power, (force-lb)(ft)/hr, (force-g)(cm)/s	FL/θ
r	Radius, ft, cm	L
C	Specific heat, Btu/(lb)(°F), cal/(g)(°C)	H/MT
v	Specific volume, ft^3/lb, cm^3/g	L^3/M
τ'	Stress, force-lb/ft^2, force-g/cm^2	F/L^2
σ	Surface tension, force-lb/ft, force-g/cm	F/L
T	Temperature, °F, °C	T
k	Thermal conductivity, Btu/(hr)(ft^2)(°F/ft), cal/(s)(cm^2)(°C/cm)	$H/LT\theta$
α	Thermal diffusivity, ft^2/hr, cm^2/s	L^2/θ
β	Thermal coefficient of expansion, 1/°F, 1/°C	1/T
R_t	Thermal resistivity, (°F)(ft)(hr)/Btu, (°C)(cm)(s)/cal	$LT\theta/H$
θ	Time, hr, s	θ
u	Velocity, ft/hr, cm/s	L/θ
μ_g	Viscosity (force-lb)(hr)/ft^2, (force-g)(s)/cm^2	$F\theta/L^2$
μ	Viscosity (abs), lb/(ft)(hr), g/(cm)(s)	$M/L\theta$
M	Mass, lb, g	M
w_0	Work, (force-lb)(ft), (force-g)(cm)	FL

F = force, H = heat, L = length, M = mass, T = temperature, θ = time

or

$$\frac{h_i}{CG} = \frac{f}{2} \qquad (3\text{-}13)$$

where G is mass velocity.

Early work conceptualized that the heat transfer through the boundary layer of thickness δ as:

$$Q = \frac{k(t' - t_p)}{\delta} \qquad (3\text{-}14)$$

where   k = mean conductivity
   t' = temperature of pipe inner circumference

For streamline flow with velocity u':

$$Q = h_i(T - t') = \frac{\tau' C(T - t')}{u - u'} \qquad (3\text{-}15)$$

This is rearranged to give:

$$h_i = \frac{Q}{(T - t_p)} = \frac{\tau'}{\dfrac{\mu u'}{k} + \dfrac{u}{C} - \dfrac{u'}{C}} \qquad (3\text{-}16)$$

For $r' = u'/u$, this becomes:

$$\frac{h}{CG} = \frac{f}{2} \frac{1}{1 - r' + r'(C/k)} \qquad (3\text{-}17)$$

This is known as the Prandtl modification of the Reynolds analogy. Note that $C\mu/k$ is the dimensionless Prandtl number.

Mean Temperature Difference

In a continuously operated system such as a heat exchanger, the temperature difference between warmer and colder fluids varies as a function of length. It is therefore, necessary to integrate the basic equation:

$$dq = UdA\Delta t_o \qquad (3\text{-}18)$$

where Δt_o is the overall temperature difference between hot and cold fluids. Principal assumptions made are constant U, constant mass flow-rates, no phase change, constant specific heats, and negligible heat losses.

For parallel or counterflow fluids:

$$q = UA\Delta t_{o,lm} \tag{3-19}$$

where the log-mean temperature difference is:

$$\Delta t_{o,lm} = \frac{\Delta t_{o1} - \Delta t_{o2}}{\ln \dfrac{\Delta t_{o1}}{\Delta t_{o2}}} \tag{3-20}$$

Δt_{o1}, Δt_{o2} are terminal temperature differences

Figure 3-2 provides a convenient nomogram for evaluating $\Delta t_{o,lm}$. The product UA is obtained from the resistances:

$$\frac{1}{UA} = \frac{1}{h_i A_i} + \frac{1}{h_{di} A_i} + \frac{x}{kA_{avg}} + \frac{1}{h_{do} A_o} + \frac{1}{h_o A_o} \tag{3-21}$$

Correlations for Heating and Cooling

The following provides useful design correlations for common situations of cooling and heating without phase changes.

Gas Flow in Long Straight Tubes

For turbulent flow, the following dimensional correlation can be used:

$$h = \frac{16.6C_p(G')^{0.8}}{(D_i')^{0.2}} \tag{3-22}$$

where C_p = specific heat of gas (Btu/lb-°F or g-cal/g-°C)
        G' = mass velocity (lb gas/s/ft²)
        D_i' = actual inside diameter (in.)

For a gas heater or cooler, the following heat balance applies:

$$q = \left(\frac{16.6C_p(G')^{0.8}}{(D_i')^{0.2}}\right)\left(\frac{\pi D_i' L}{12}\right)(\Delta t)_m \tag{3-23}$$

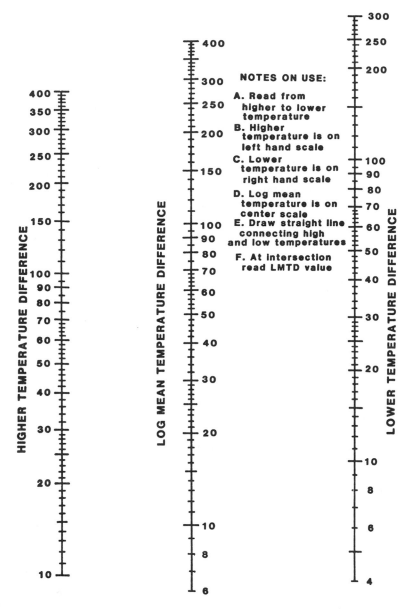

NOTES ON USE:

A. Read from higher to lower temperature

B. Higher temperature is on left hand scale

C. Lower temperature is on right hand scale

D. Log mean temperature is on center scale

E. Draw straight line connecting high and low temperatures

F. At intersection read LMTD value

HIGHER TEMPERATURE DIFFERENCE

LOG MEAN TEMPERATURE DIFFERENCE

LOWER TEMPERATURE DIFFERENCE

Figure 3-2. Nomogram for evaluating log-mean temperature difference.

and

$$\left(\frac{L}{D_i'}\right)\left[\frac{\Delta t_m}{(t_1 - t_2)}\right] = 4.52(D_i'G')^{0.2} \tag{3-24}$$

where t_i, t_2 = lower and higher gas temperatures, respectively
t_m = logarithmic-mean temperature difference between gas and tube surface (°F)
L = heated length (ft)

Gas Flow in Coiled Tubes

The following is applicable to small tubes (up to $1^1/4$-in. steel pipe) for turbulent Reynolds numbers up to 150,000:

$$\frac{hD}{k} = \left(0.039 + \frac{0.138D}{D_c}\right)\left(\frac{DC_pG}{k}\right)^{0.76} \tag{3-25}$$

where D/D_c is the ratio of the diameter of the tube to that of the helix.

High Velocity Gas Flows

For Mach numbers in the range of $0.2 \sim 1.0$, through a long straight tube, h can be computed from the following dimensionless correlation:

$$\left(\frac{h}{C_pG}\right)\left(\frac{C_p\mu}{k}\right)^{2/3} = \frac{0.027}{(DG/\mu_m)^{0.23}} \tag{3-26}$$

where μ_m is the mean fluid viscosity

Low Viscosity Liquids Inside Tubes

For turbulent flows (Re > 2,100) the following dimensionless correlation can be used:

$$\frac{h}{C_pG}\left(\frac{C_p\mu_f}{k}\right)^{2/3} = 0.023\left(\frac{DG}{\mu_f}\right)^{-0.2} = \frac{f}{2} \tag{3-27}$$

High Viscosity Liquids in Streamline Flow

For high viscosity liquids (e.g., viscous oils) in laminar flow (Re < 2,100), the Sieder-Tate equation can be used:

$$\frac{h_{am}D}{k} = 2.0 \left(\frac{WC_p}{kL}\right)^{1/3} \left(\frac{\mu'}{\mu_s}\right)^{0.14} \tag{3-28}$$

h_{am} is based on the arithmetic mean of the terminal temperature differences.

Equation 3-28 can also be applied for Re > 2,100, but calculations must be based on the logarithmic mean. Note W = mass flowrate per tube of heated length.

For cases where the Grashof number (Gr = $D^3\rho^2gL\beta\Delta t/\mu^2$, evaluated at the stream bulk temperature) is large and Re is small, the right-hand side of Equation 3-28 should be multiplied by the following dimensionless group for horizontal tube flow:

$$C' = \frac{2.3(1 + 0.01Gr^{1/3})}{\log Re} \tag{3-29}$$

Liquid Flow in Vertical Pipes

For turbulent, vertical flow in pipes:

$$\frac{h_m}{[k_f^2\rho_f^2C_pg\beta(\Delta t)/\mu_f]^{1/3}} = 0.13 \tag{3-30}$$

Equation 3-30 is based on water flow data only;

where k = thermal conductivity
  C_p = specific heat at constant pressure
  ρ = density
  β = coefficient of expansion
  Δt = temperature difference between wall and fluid
  g = gravitational constant
  μ = viscosity

Gas Flows at Right Angles to Tubes

For air flow at right angles to the axes of single cylinders ranging in diameter from $0.001 \sim 3.75$ in., and for gas temperatures up to $1800°F$, the following correlations apply:

$$\frac{h_m D_o/k_f}{(C_p\mu_f/k_f)^{0.3}} = 0.35$$

$$+ 0.47 \left(\frac{D_o G}{\mu_f}\right)^{0.52}; \ (0.1 < Re < 1,000) \qquad (3\text{-}31)$$

$$\frac{h_m D_o/k_f}{(C_p\mu_f/k_f)^{0.3}} = 0.26 \left(\frac{D_o G}{\mu_f}\right)^{0.6}; \ (1,000 < Re < 50,000) \qquad (3\text{-}32)$$

For gases at moderate temperatures at $1,000 < Re < 50,000$:

$$h = \frac{0.3 C_p G^{0.6}}{(D_o)^{0.4}} \qquad (3\text{-}33)$$

For gases flowing at right angles to a bank of staggered tubes $(3,000 < Re < 40,000)$:

$$\left(\frac{h_m}{G_p G_{max}}\right)\left(\frac{C_p\mu_f}{k}\right)^{2/3} = 0.33\left(\frac{D_o G_{max}}{\mu_f}\right)^{-0.4} \qquad (3\text{-}34)$$

Liquid Flows Normal to Tube Banks

For water flows at right angles to a bank of staggered tubes, the following correlation can be used for the mean heat transfer coefficient.

$$h_m = \frac{370(1 + 0.0067t_f)(u_s^{0.6})}{(D_o')^{0.4}} \qquad (3\text{-}35)$$

where u_s = average velocity (volumetric rate divided by cross-section filled by fluid, fps)

$t_f = (t + t_w)/2$

Natural Convection

For free or natural convection of heat between a solid and fluid, without a change in state, the following dimensionless correlation applies:

$$\frac{h_c D}{k_f} = \frac{1}{a}\left(\frac{D^3 \rho_f^2 g\beta\Delta t}{\mu_f^2}\right)^n \left(\frac{C_p \mu_f}{k_f}\right)^m \tag{3-36}$$

For pipes having large length-to-diameter ratios, use the following values:

$a = 2.12$
$n = 0.27$
$m = 0.25$

Flow Over Miscellaneous Surfaces

For heat loss by conduction and convection to air at atmospheric pressure and ambient temperatures, the following dimensional correlations can be used for estimates of h:

Horizontal and long vertical pipes:

$$h_c = 0.5(\Delta t_s / D_o')^{0.25} \tag{3-37}$$

Vertical plates less than 2 ft high:

$$h_c = 0.28(\Delta t_s / H)^{0.25} \tag{3-38}$$

Vertical plates more than 3 ft high:

$$h_c = 0.27(\Delta t_s)^{0.25} \tag{3-39}$$

Horizontal plates facing upwards:

$$h_c = 0.38\Delta t_s^{0.25} \tag{3-40}$$

Horizontal plates; facing downward:

$$h_c = 0.2\Delta t_s^{0.25} \tag{3-41}$$

where Δt_s = temperature of exposed surface minus ambient temperature (°F)
  H = plate height (ft)
  D' = inside diameter (in.)
  h_c = Btu/hr-ft²-°F

Laminar Flow Heat Transfer in Annuli

The empirical correlation of Chen et al.[1] can be used for laminar flow heat transfer. The correlation is based on heat transfer data from the inner wall in four annuli with diameter ratio d_2/d_1 in the range of $1.09 \sim 2$.

$$\mathrm{Nu}_{am} = 1.02(\mathrm{Re})^{0.45}(\mathrm{Pr})^{0.5}\left(\frac{\mu_b}{\mu_1}\right)_{am}^{0.14}\left(\frac{d_e}{L}\right)^{0.4}\left(\frac{d_2}{d_1}\right)^{0.8}(\mathrm{Gr})^{0.05} \quad (3\text{-}42)$$

where μ_b = viscosity at the arithmetic mean bulk temperature
μ_1 = viscosity at temperature T_1 of the inner wall of the annulus

Other physical properties are based on the fluid bulk temperature. Equation 3-42 is applicable for $200 < \mathrm{Re} < 2,000$, where Re is based on the equivalent diameter d_e.

Notes on Correlating Heat Transfer Data

For turbulent flows, heat transfer data are correlated in terms of dimensionless groups. The most commonly used groups are Nusselt, Reynolds, and Prandtl numbers. Other groups such as the Peclet and Stanton numbers are derived from the above three groups.

The two classical forms of equations relating these dimensionless groups are:

$$\mathrm{Nu} = C_1(\mathrm{Re})^{n_1}(\mathrm{Pr})^{n_2}\left(\frac{d_w}{L}\right)^{n_3} \quad (3\text{-}43)$$

$$\mathrm{St} = C_2(\mathrm{Re})^{n_4}(\mathrm{Pr})^{n_5}\left(\frac{d_w}{L}\right)^{n_6} \quad (3\text{-}44)$$

where C_1, C_2 are constants; $n_1 \ldots n_6$ = exponents.

Both equations are equivalent, each having certain advantages.

In terms of the definitions of Pr and Re the equations for tube flow become:

$$\mathrm{Nu}_m = \frac{d_w^2}{4L}\frac{GC_p}{k}\frac{T_{b2} - T_{b1}}{(T_w - T_b)_{lm}} \quad (3\text{-}45)$$

$$St_m = \frac{d_w}{4L} \frac{T_{b2} - T_{b1}}{(T_w - T_b)_{lm}} \tag{3-46}$$

where T_{b1}, T_{b2} are the inlet and outlet bulk temperatures, respectively

Fully Developed Turbulent Flow

The Dittus-Boelter equation[2] is applicable for moderate temperature differences:

$$\left(\frac{h_m d_w}{k}\right)_b = 0.023\left(\frac{d_w G}{\mu}\right)_b^{0.8} \left(\frac{C_p \mu}{k}\right)_b^n \tag{3-47}$$

Equation 3-47 is based on fluid properties evaluated at the arithmetic mean bulk temperature.

Exponent $n = 0.4$ for heating; $= 0.3$ for cooling. The correlation is applicable over the range:

Re $> 10,000$
$0.7 < $ Pr < 100
$L/d_w > 60$

In terms of the Stanton number, the Colburn[3] correlation (Equation 3-27) applies.

Turbulent Flow Heat Transfer in Annuli

The following correlation [4] is based on heat transfer data for flow in annuli:

$$\left(\frac{h_m d_e}{k}\right)_b = 0.023 \left(\frac{d_e G}{\mu}\right)_b^{0.8} \left(\frac{C_p \mu}{k}\right)_b^{0.4} \left(\frac{d_2}{d_1}\right)^{0.45} \tag{3-48}$$

The correlation applies for Re $> 10^4$ and $d_e = d_2 = d_1$. Properties are based on bulk temperatures.

Flow in Noncircular Smooth Ducts

For air flowing in square, rectangular, and triangular ducts, the following correlation is useful:[5]

$$\left(\frac{h_m d_e}{k}\right) = 0.023 \left(\frac{d_e G}{\mu}\right)^{0.8} \left(\frac{C_p \mu}{k}\right)^{0.4} \tag{3-49}$$

In addition to straight dimensional analysis, analogies between momentum and heat transfer have enabled semi-empirical models to be developed. One such analogy was noted earlier (Prandtl/Reynolds analogy). Other important analogies are that of von Karman,[6] Colburn,[3] and Martinelli.[7] The following notes apply to fully developed turbulent pipe flows.

von Karman Analogy

This is a modification of Prandtl's analogy, and makes use of the universal velocity distribution:

$$Nu = \frac{(f/2)\ RePr}{1 + 5\sqrt{f/2}\{Pr\text{-}1 + \ln\ [1 + 5/6(Pr\text{-}1)]\}} \tag{3-50}$$

Colburn Analogy

This is based on the j-factor, derived from Equation 3-27:

$$\left(\frac{h_m}{GC_p}\right)_b \left(\frac{C_p\mu}{k}\right)^{2/3} f/2 = j_H \tag{3-51}$$

where j_H is the j-factor for heat transfer.
Since $St = Nu/RePr$, then

$$Nu = \frac{f}{2}\ Re(Pr)^{1/3} \tag{3-52}$$

And for mass transfer:

$$\frac{k_w d_w}{D} = \frac{f}{2}\ Re(Sc)^{1/3} \tag{3-53}$$

where Sc is the Schmidt number and D is diffusivity.

Martinelli Analogy

The analysis is based on:

1. Fully developed velocity and temperature profiles
2. Constant fluid properties
3. Uniform heat flux along the tube wall. The principle equation developed describes the temperature distribution in the fluid over a laminar sublayer, buffer layer, and turbulent core and is described

by Knudsen.[8] A similar analysis was presented by Deissler[9,10] and is described and applied to film flow by Cheremisinoff.[11]
Additional correlations and alternative methods are outined in the sample calculations.

Sample Calculation 3-1. (Evaluation of Heat Transfer Coefficient). Estimate the heat transfer coefficient for steam condensation on the walls of a 2-m long pipe inclined from the horizontal by an angle of 60°. The wall temperature is 210°F and the steam temperature is 220°C.

Solution.

• First obtain physical properties data using an average temperature

$$t_{av} = \frac{t_w + t}{2} = \frac{210 + 220}{2} = 215°C$$

From appropriate physical properties tables:

Conductivity k = 0.56 kcal/m-hr-°C = 0.625 Wt/m-°C
Viscosity μ = 1.31 × 10^{-5}kg$_f$-s/m^2 = 1.29
    × 10^{-4}N-s/m^2
Density ρ = 846 kg/m^3
Heat of vaporization λ = 449 kcal/kg = 1.88 × 10^6J/kg

• The following correlations may be used to evaluate the heat transfer coefficient:

For laminar film flow:

$$h = \frac{h'}{\sqrt[4]{H(t - t_w)}} \qquad (3\text{-}54)$$

where

$$h' = 0.943 \sqrt[4]{\frac{\lambda \rho^3 g k^3}{\mu}} \qquad (3\text{-}55)$$

For turbulent film flow:

$$h = 3 \times 10^{-3} \sqrt{\frac{H(t - t_w)k^3 \rho^2 g}{\lambda \mu^3}} \qquad (3\text{-}56)$$

where H = enthalpy

λ = heat of vaporization

To evaluate which formula to use:

$$H(t - t_w)_{cr} = \frac{2,680\lambda\mu^{5/3}}{k\rho^{2/3}g^{1/3}}$$

$$= 2,680 \frac{1.88 \times 10^6(1.29 \times 10^4)^{5/3}}{0.652 \times 846^{2/3} \times 9.81^{1/3}}$$

$$= 13.3\text{m-}°C$$

The criterion for turbulent film flow is $H(t - t_w) > 13.3$. And since $H(t - t_w) = 2(220 - 210) > 13.3$, use Equation 3-56.

$$h = 3 \times 10^{-3} \sqrt{\frac{H(t - t_w)k^3\rho^2g}{\lambda\mu^3}}$$

$$= 3 \times 10^{-3} \sqrt{\frac{2 \times 10 \times 0.652^3 \times 846^2 \times 9.81}{1.88 \times 10^6(1.29 \times 10^{-4})^3}}$$

$$= 9,330 \text{ J/m}^2\text{-s-}°C$$

The heat transfer coefficient for the inclined pipe is:

$$h_\beta = h\sqrt[4]{\sin \beta}$$

$$= 9,330 \sqrt[4]{\sin 60°}$$

$$= 9,050 \text{ J/m}^2\text{-s-}°C \text{ (or } 7,750 \text{ kcal/m}^2\text{-hr-}°C)$$

Sample Calculation 3-2. (Heat Transfer Coefficient for Boiling Solution). A 40% solution of NaOH is inside a pipe of the following dimensions: diameter = 57 × 50 mm, length = 2.8 m. The pressure and temperature are 18,400 N/m² and 85°C, respectively. The solution is brought to boiling conditions using saturated steam at 112°C. Estimate the heat transfer coefficient.

Solution.

- The following correlation can be used to evaluate the heat transfer coefficient:

$$h = h_w \left(\frac{k}{k_w}\right)^{0.565} \left[\left(\frac{\rho}{\rho_w}\right)^2 \left(\frac{C}{C_w}\right)\left(\frac{\mu_w}{\mu}\right)\right]^{0.435} \tag{3-57}$$

Subscript w refers to water.

- The coefficient for pure water can be computed from:

$$h_w = 0.22P^{0.58}(t_w - t_k)^{2.33}, \text{ kcal/m}^2\text{-hr-}°C \tag{3-58}$$

or

$$h_w = 25.6P^{0.58}(t_w - t_k)^{2.33}, \text{ Wt/m}^2\text{-}°C \tag{3-59}$$

- Physical properties are gathered from the literature:

Property	H_2O	NaOH Solution
Viscosity $\mu \times 10^3$, N-s/m^2	0.336	2.92
Specific heat, J/kg-°C	4190	3445
Density, kg/m^3	969	1390
Conductivity, Wt/m-°C	0.678	0.56

- Using the preceding formulas:

$$h = 25.6\left(\frac{18,400}{9.81 \times 10^4}\right)^{0.58}(t_w - 85)^{2.33}\left(\frac{0.56}{0.678}\right)^{0.565}$$

$$\times \left[\left(\frac{1,390}{969}\right)^2 \times \frac{3,440}{4,190} \times \frac{0.336}{2.92}\right]^{0.435}$$

$$= 4.21(t_w - 85)^{2.33} \text{ Wt/m}^2\text{-}°C$$

- It is useful to prepare a sketch of the system as shown in Figure 3-3. Note that the temperature of the inside pipe wall is not known. Assuming the film to be flowing downward in the laminar regime,

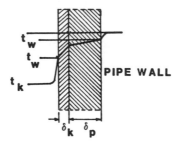

Figure 3-3. System for Sample Calculation 3-2.

the heat transfer coefficient at steam condensation can be computed from the following formula:

$$h = \frac{h'}{\sqrt[4]{H(t - t_w')}}$$

$$h = \frac{12,658}{\sqrt[4]{2.5(112 - t_w)}}$$

The temperature of the inside pipe wall and the heat transfer coefficient can be evaluated from the following formula:

$$h_{in}r_{in}(t_w - t_k) = \frac{t_w' - t_w}{\dfrac{\delta_k}{k_k r_{av,k}} + \dfrac{\delta_p}{k_p r_{av,p}}}$$

$$= h_{out}r_o(t - t_w') = h'(t - t_w)$$

where r_{in} = internal pipe radius = $0.050/2 - \delta_k$ (determined as difference between inside pipe radius and thickness of the layer of formed crystals) (m)

$r_{av,k}$ = average radius of crystal layer on the inside pipe surface (assumed to be equal to inside pipe radius)

$r_{av,p}$ = $(0.050 + \delta_p)/2$ – average pipe radius

r_{out} = $0.057/2$ – outside pipe radius (m)

δ_k = 0.5 – thickness of crystal layer (mm)

$$\delta_p = (0.057 - 0.05)/2 = 0.0035 - \text{pipe wall thickness, (m)}$$

$$k = 1.5 - \text{conductivity of crystal layer (Wt/m-°C)}$$

$$k_p = 45 - \text{conductivity of steel (Wt/m-°C)}$$

• Substituting values into the previous equations:

$$4.21(t_w - 85)^{2.33}\left(\frac{0.050}{2} - 0.0005\right)(t_w - 85)$$

$$= \frac{t'_w - t_w}{\dfrac{0.0005}{1.5 \times 0.025} + \dfrac{0.0035}{45 \times 0.02675}} = 10,000\,(111.7 - t'_w)^{-1.4}$$

$$\times \frac{0.057}{2} \times (11.7 - t'_w) = h'\frac{0.057}{2}\,(111.7 - 85)$$

$$0.103(t'_w - 85)^{3.33} = 61.5(t'_w - t_w) \tag{3-60}$$

$$0.103(t_w - 85)^{3.33} = 285(11.7 - t'_w)^{3/4} \tag{3-61}$$

$$0.103(t_w - 85)^{3.33} = 0.76h' \tag{3-62}$$

The intersection of the correlations gives the solution $h' = 829$ Wt/m²-°C and the internal wall temperature is $t_w = 99\,°C$.

Sample Calculation 3-3. (Heat Losses Through a Furnace Wall). A single-walled furnace has a thickness of 0.6 m. The mean temperature inside the furnace is 850°C (t_g) and the mean ambient air temperature is 28°C (t_a). The heat transfer coefficient from the gases to the wall is $h_g = 32$ Wt/m²-°C, and the heat transfer coefficient from the wall to the air is $h_a = 12$ Wt/m²-C°. The average wall conductivity is $k = 1.8$ Wt/m-°C. Estimate the outside wall temperature and heat losses through the furnace wall.

Solution.

• Losses can be computed from:

$$Q/A = h\Delta t$$

where the overall heat transfer coefficient is computed from:

$$h = \cfrac{1}{\cfrac{1}{h_g} + \cfrac{\delta}{k} + \cfrac{1}{h_a}}$$

$$= \cfrac{1}{\cfrac{1}{32} + \cfrac{0.6}{1.8} + \cfrac{1}{12}} = 2.23 \ Wt/m^2$$

• Heat losses are:

$$Q/A = 2.23(850 - 28) = 1833 \ Wt/m^2$$

• To compute the outer wall temperature we can assume that the same amount of heat is transferred to the air from the outer wall surface.

$$Q/A = h_a(t_o - t_{in}) = 12(t_o - 28) = 1{,}833$$

or

$$t_o = 181°C.$$

Sample Calculation 3-4. (Furnace Wall Temperature Control). Continuing with the previous example, it is decided that the outer wall temperature is too high. Two possible approaches to lowering this temperature are to add a layer of insulation to the outer wall or to enhance the rate of heat transfer from the surface by forced convection. Evaluate the thickness of insulation required (assume diatome, $k = 0.13 \ Wt/m-°C$) and the air velocity required to maintain an outside wall temperature 130°C.

Solution.

• First evaluate the required insulation thickness: The heat loss under the new conditions are:

$$Q/A = h_a(t_o - t_{air}) = 12(130 - 28) = 1{,}244 \ Wt/m^2$$

The required insulation thickness can be solved from:

$$\frac{Q}{A} = \frac{t_g - t_o}{\dfrac{1}{h_g} + \dfrac{\delta}{k} + \dfrac{\delta_{ins}}{k_{ins}}} \tag{3-63}$$

$$1{,}224 = \frac{850 - 130}{\dfrac{1}{32} + \dfrac{0.6}{1.8} + \dfrac{\delta_{ins}}{0.13}}$$

$$\delta_{ins} = 0.03 \text{ m (or 1.1 in.)}$$

- Next evaluate the air flow velocity required to maintain the same outer wall temperature:

Heat losses are:

$$\frac{Q}{A} = \frac{t_g - t_o}{\dfrac{1}{h_g} + \dfrac{\delta}{k}} = \frac{850 - 130}{\dfrac{1}{32} + \dfrac{0.6}{1.8}} = 1{,}975 \text{ Wt/m}^2$$

The heat transfer coefficient from the wall to air is:

$$Q/A = h_a(t_o - t_a)$$
$$1{,}975 = h_a(130 - 28)$$
$$h_a = 19.4 \text{ Wt/m}^2\text{-}^\circ\text{C}$$

The required air velocity can be estimated from the following Nusselt number expression:

$$Nu = 0.057(RePr)^{0.78}$$

Note: $Re = \dfrac{uL}{\nu}$

$$u = \frac{\nu}{L} \frac{1}{Pr} \left(\frac{Nu}{0.057}\right)^{1/0.78} \tag{3-64}$$

Assuming an average temperature, $t_{av} = (t_o + t_a)/2 = (130 + 28)/2 = 79^\circ\text{C}$, physical properties of air are obtained from the literature:

Kinematic viscosity $\nu = 0.205 \times 10^{-4} m^2/s$
Conductivity $k = 2.46 \times 10^{-2} kcal/m\text{-}hr\text{-}°C$ (or 2.78 $Wt/m\text{-}°C$)

Hence, the Prandtl number is Pr = 0.722. And the Nusselt number:

$$Nu = \frac{hL}{k} = \frac{19.4 \times 2}{2.46 \times 10^{-2}} = 1,577$$

Solving for air velocity u:

$$u = \frac{0.205 \times 10^{-4}}{2} \times \frac{1}{0.722} \left(\frac{1,577}{0.057}\right)^{1/0.78} = 7.03 \text{ m/s.}$$

Note that Equation 3-64 is only valid for Re $> 5 \times 10^5$, so the Reynolds number should be checked:

$$Re = \frac{uL}{\nu} = \frac{7.03 \times 2}{2.05 \times 10^{-5}} = 6.859 \times 10^5$$

Hence the calculation is valid.

Sample Calculation 3-5. (Heat Flow Through a Reactor Wall). Estimate the amount of heat transferred through a reactor wall that is 0.3 m thick. The wall's conductivity varies with temperature and can be described by the following correlation:

$$k = 0.35(1 + 0.7 \times 10^{-3}t), \text{ Wt/m}^2\text{-}°C$$

The temperature of the vapors inside the reactor is 912°C, and the outside air temperature is 35°C. The heat transfer coefficients from vapors to wall and from wall to air are: $h_g = 21$ Wt/m²-°C and $h_a = 12$ Wt/m²-°C, respectively. The surface area of the reactor is (assume a flat wall for illustration), is 70 m².

Solution.

• The following equation applies:

$$Q = \frac{A(t_g - t_a)}{\dfrac{1}{h_g} + \dfrac{\delta}{k_{av}} + \dfrac{1}{h_a}} \tag{3-65}$$

Estimate the wall conductivity using an average temperature:

$$k_{av} = 0.35 \left(1 + 0.7 \times 10^{-3} \frac{t_1 + t_2}{2}\right)$$

Wall temperatures t_1 and t_2 can be expressed in terms of heat losses:

$$Q = h_g A(t_g - t_1) = h_a A(t_2 - t_a)$$

$$t_1 = t_g - \frac{Q_\ell}{h_g A}; \quad t_2 = t_a + \frac{Q_\ell}{h_a A}$$

- Substituting values into Equation 3-65:

$$Q_\ell = \frac{A(t_g - t_a)}{\dfrac{1}{h_g} + \dfrac{0.30}{0.35\left[1 + \dfrac{0.7 \times 10^{-3}}{2}\left(t_g - \dfrac{Q_\ell}{h_g A} + t_a + \dfrac{Q_\ell}{h_a A}\right)\right]} + \dfrac{1}{h_a}}$$

$$Q_\ell = \frac{70(912 - 35)}{\dfrac{1}{21} + \dfrac{0.30}{0.35\left[1 + 3.5 \times 10^{-4}\left(912 - \dfrac{Q_\ell}{21 \times 70} + 35 + \dfrac{Q_\ell}{12 \times 70}\right)\right]} + \dfrac{1}{12}}$$

Solving this expression for the heat transferred:

$$Q = 16{,}937 \text{ Wt}$$

The wall temperatures can now be computed:

$$t_1 = 912 - \frac{16{,}937}{21 \times 70} = 900°C$$

$$t_2 = 35 + \frac{16{,}937}{12 \times 70} = 55°C$$

Sample Calculation 3-6. (Heat Loss from Molten Metal). Molten lead is circulated at a rate of 25 kg/hr in a cylindrical vessel that is 15-m tall with an outside diameter of 35 mm and an inner diameter of 30 mm. The vessel is insulated with a material having a thermal conductivity of

0.15 Wt/m-°C. The heat transfer coefficient from the molten metal is $h_1 = 3,200$ Wt/m²-°C, and from vessel to air $h_2 = 12$ Wt/m²-°C. The conductivity of the vessel is $k_s = 45$ Wt/m²-°C; the specific heat of the lead is $C = 0.0375$ kcal/kg-°C; the air temperature $t_a = 23$°C. Determine the insulation thickness required to prevent the lead from cooling by no more than 5°C. The temperature of the lead is 400°C.

Solution.

- The allowable heat losses are:

$$Q_\ell = G \times C(t_{in} - t_{fin})$$

$$Q_\ell = 25,000 \times 0.0375 \times 5 = 4,688 \text{ kcal/hr}$$

$$= Q_\ell = 4,688 \frac{4.19 \times 10^3}{3,600} = 5,456 \text{ Wt}$$

- The insulation thickness is determined from:

$$Q_\ell = \frac{2}{\dfrac{1}{h_1 r_{in}} + \dfrac{\delta_p}{k_\ell r_{af,p}} + \dfrac{\delta_{ins}}{k_{ins} r_{av,ins}} + \dfrac{1}{h_2 r_{out}}} \qquad (3\text{-}66)$$

where Δt = temperature difference between average lead temperature $(400 + 400 - 5)/2 = 397.5$°C and the ambient temperature

r_{in} = inside pipe radius = $30/2$ mm = 0.015 m

$r_{av,p}$ = mean logarithmic pipe radius, which may be approximated by the mean arithmetic radius when the wall thickness is small

$$r_{av,p} = r_{in} + \frac{\delta_p}{2} = 0.015 + \frac{0.0035}{2} = 0.0168 \text{ m}$$

$r_{av,ins}$ = mean logarithmic radius of insulation

$$r_{av,ins} = \frac{0.0175 + \delta_{ins} - .0175}{\ln \dfrac{.0175 + \delta_{ins}}{.0175}}$$

r_{out} = outside radius

$$r_{out} = r_{in} + \delta_p + \delta_{ins} = 0.0175 + \delta_{ins}$$

- Substituting values into Equation 3-66 (assuming an ambient air temperature of 20°C):

$$Q_\ell = \cfrac{(2)(15)(397.5 - 20)}{\cfrac{1}{3,200 \times 0.0175} + \cfrac{0.0035}{45 \times 0.0168} + \cfrac{\delta_{ins}}{0.15\delta_{ins}} + \cfrac{1}{12(0.0175 + \delta_{ins})}}$$

$$= 5,456$$

Solving for δ_{ins}:

$$\ln\left(\frac{0.0175 - \delta_{ins}}{0.0175}\right) = 0.97 - \frac{0.0128}{0.0175 + \delta_{ins}}$$

$$\delta_{ins} = 0.0127 \text{ m (or 0.5 in.)}$$

Sample Calculation 3-7. (Thermocouple Correction). A thermocouple is being used to sense the temperature of gas inside a reactor (350°C is detected). The thermocouple is installed through a metal case that passes through the wall of the reactor as shown in Figure 3-4. The casing diameter is 0.0075 m, and its length is 0.10 m. The thickness of the casing is 0.0015 m; its conductivity k = 50 Wt/m-hr, and h = 8 Wt/m²-°C. The temperature of the reactor wall is independently measured to be 110°C. Evaluate whether this thermocouple arrangement will provide an accurate measurement of the gas temperature.

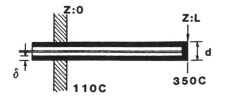

Figure 3-4. Thermocouple set-up in Sample Calculation 3-7.

Solution. The set-up shown in Figure 3-4 cannot possibly sense the true temperature inside the reactor. In reality, it senses the temperature of the internal end of the metal case. The problem, therefore, is to determine how good an approximation is obtained.

We must evaluate the relationship between these temperatures:

- The heat balance at distance Z from the wall:

$$\frac{d^2\Delta t}{dZ^2} = m^2\Delta t \tag{3-67}$$

where $m^2 = \dfrac{h\Gamma}{kS}$

$\Delta t = t - t_g$, difference between temperature of case wall and gas temperature

Γ = wetted perimeter (outer perimeter)

S = case cross-sectional area

- Solve the following differential equation derived from heat balance:

$$\Delta t = C_1 e^{mZ} + C_2 e^{-mZ} \tag{3-68}$$

where C_1, C_2 are constants evaluated from the boundary conditions.

At $Z = 0$, temperature in this case equals the wall temperature, hence,

$$\Delta t_o = C_1 + C_2$$

- To obtain the boundary condition at $Z = L$, note that the heat flux due to conduction for $Z = L$ and the heat absorbed by the internal end of the case can be expressed as:

$$-kS\left(\frac{dt}{dz}\right)_{Z=L} = h\frac{\pi d^2}{4}(\Delta t)_{Z=L}$$

where $(\Delta t)_{Z=L} = C_1 e^{mL} + C_2 e^{-mL}$

$$\left(\frac{dt}{dz}\right)_{Z=L} = \left(\frac{d\Delta t}{dz}\right)_{Z=L} = C_1 m e^{-mL} - C_2 m e^{-mL}$$

- To evaluate the integration constants, the second equation becomes:

$$-kS(C_1me^{mL} - C_2me^{-mL}) = h\frac{\pi d^2}{4}(C_1e^{mL} - C_2e^{-mL})$$

- Evaluating the constants of integration gives:

$$C_1 = \frac{\Delta t_o}{1 + \dfrac{\theta + 1}{\theta - 1}e^{2mL}}$$

$$C_2 = \frac{\Delta t_o \dfrac{\theta + 1}{\theta - 1}e^{2mL}}{1 + \dfrac{\theta + 1}{\theta - 1}e^{2mL}}$$

where $\theta = m\dfrac{4kS}{\pi hd^2}$

- The temperature difference at $Z = L$ is:

$$(\Delta t)_{Z = L} = \Delta t_o e^{mL} \frac{1 + \dfrac{\theta + 1}{\theta - 1}}{1 + \dfrac{\theta + 1}{\theta - 1}e^{2mL}}$$

- Evaluating m and θ:

$$S = \frac{\pi}{4}[d^2 - (d - 2\delta)^2]$$

$$= \frac{\pi}{4}[0.0075^2 - (0.0075 - 2 \times .0015)^2]$$

$$S = 2.83 \times 10^{-5} \text{ m}^2$$

$$\Gamma = \pi d = \pi(0.0075) = 2.36 \times 10^{-2} \text{ m}$$

$$m = \sqrt{\frac{8 \times 2.36 \times 10^{-2}}{50 \times 2.83 \times 10^{-5}}} = 11.6 \text{ m}^{-1}$$

$$\theta = 11.6\frac{4(50)(2.83 \times 10^{-5})}{\pi(8)(0.0075)^2} = 46.3$$

- Evaluating the actual gas temperature:

$$(\Delta t)_{Z = L} = (100 - t_g)e^{11.6 \times 0.10} \left(\frac{1 + \dfrac{46.3 + 1}{46.3 - 1}}{1 + \dfrac{46.3 + 1}{46.3 - 1}\, e^{2 \times 11.6 \times 0.10}} \right)$$

$$= 350 - t_g$$

$$t_g = 657°C$$

- The previous calculations show that the thermocouple arrangement of Figure 3-4 reports a temperature over 300°C lower than the actual temperature inside the reactor.

    To reduce this error, the casing should be insulated, the casing constructed from a low conductivity material, and the wall of the casing should be ultra-thin to provide a high heat transfer coefficient from gas to thermocouple.

Sample Calculation 3-8. (Heat Losses from Steam Line). Estimate the heat losses from a vertical steam line that is 5-m long and 180-mm in diameter. The wall temperature is 185°C and the ambient air temperature is 5°C.

Solution.

- First compute the average temperature:

$$t_{av} = \frac{1}{2}(t_w + t_a) = \frac{185 + 5}{2} = 95°C$$

- The following data is obtained:

$$k_{av} = 0.032 \text{ kcal/m-hr-°C}$$
$$\nu_{av} = 2.35 \times 10^{-5} \text{m}^2/\text{s}$$
$$Pr_{av} = 0.70$$

- Compute the Grashoff number:

$$Gr_{av} = \frac{\beta d^3 g t}{\nu^2} = \frac{1}{368} \times \frac{(0.18)^3 \times 9.81 \times 180}{(2.35)^2} \times 10^{10}$$

$$= 5.07 \times 10^7$$

• Compute the Nusselt number:

$$(Gr \times Pr)_{av} = 5.07 \times 10^7 \times 0.70 = 3.55 \times 10^7$$
$$C = 0.54 \text{ and } n = 1/4$$

Hence:

$$Nu_{av} = 0.54 (GrPr)_{av}^{1/4} = 0.54(3.55 \times 10^7)^{0.25}$$

$$Nu_{av} = 41.7$$

• Compute the heat transfer coefficient:

$$h = \frac{Nu_{av}k_{av}}{d} = \frac{41.7 \times 0.032}{0.180} = 7.41 \frac{kcal}{m^2\text{-hr-}°C}$$

• Heat losses are:

$$Q = hA(t_w - t_a)$$

$$A = \frac{1}{4}\pi d^2 L = \frac{\pi}{4}(0.180)^2 \, 5 = 0.125$$

$$Q = 7.41(0.125)(180) = 1,667 \text{ kcal/hr}$$

Sample Calculation 3-9. (Estimating Heat Transfer Coefficient). Air flows through an 85-mm diameter duct that is 5 m long. The average air velocity is 3 m/s. Estimate the heat transfer coefficient for an average air temperature of 95°C.

Solution.

• From literature, at 95°C:

$$k_f = 0.0276 \text{ Kcal/m-hr-}°C$$

$$\nu_f = 23.1 \times 10^{-6} \text{ m}^2/s$$

• Compute the Reynolds number:

$$Re_f = \frac{ud}{\nu} = \frac{3 \times 0.085}{23.1 \times 10^{-6}} = 11,039$$

$Re_f^{0.8} = 1,715$

Prandtl number for air $Pr \simeq 0.70$

- Compute Nusselt number:

$Nu = 0.018\, Re_f^{0.8} = 0.018 \times 1,715 = 30.9$

- Heat transfer coefficient:

$$h = Nu_f \frac{k_f}{d} = 30.9 \left(\frac{0.0276}{0.085} \right) = 10.0 \ kcal/m^2\text{-}hr\text{-}°C$$

Sample Calculation 3-10. Estimate the heat transfer coefficient for methyl alcohol flowing in the intertubular space of a shell-and-tube heat exchanger with baffles. The outer tube diameter is 31 mm. The mass flow rate is 190 kg/m^2-s, and the average temperature is 38°C. Physical properties are as follows:

Viscosity $\mu = 0.465$ cp (0.465×10^{-3} N-s/m^2)
Conductivity $k = 0.179$ kcal/m-hr-°C (0.208 Wt/m-°C)
Specific heat $C = 0.63$ kcal/kg-°C (2,660 J/kg-°C)

Solution.

$$Re = \frac{uL}{\mu} = \frac{190 \times .031}{0.465 \times 10^{-3}} = 12,667$$

$$Pr = \frac{\mu C}{k} = \frac{(0.465 \times 10^{-3})(2,660)}{0.208} = 5.9$$

$$Nu = 0.2 Re^{0.6} Pr^{0.33}$$
$$= 0.2(12,667)^{0.6}(5.9)^{0.33} = 104$$

$$h = \frac{k}{L} Nu = \frac{0.208}{0.031}(104) = 698 \ Wt/m^2\text{-}°C$$

References

1. Chen, C. Y., G. A. Hawkins, and H. L. Solberg, *Trans. ASME,* 68:99 (1946).
2. Dittus, F. W., and L. M. K. Boelter, Univ. Calif. (Berkeley) *Publs. Eng.,* 2:443 (1930).
3. Colburn, A. P., *Trans. AIChE,* 29:174 (1933).
4. Wiegand, J. H., *Trans. AIChE,* 41:147 (1945).
5. Lowdermilk, W. H., W. F. Weiland, and J. N. B. Livingood, NACA RM E53J07 (1954).
6. von Karman, T., *Trans. ASME,* 61:705 (1939).
7. Martinelli, R. C., *Trans. ASME,* 69:947 (1947).
8. Knudsen, J. G., and D. L. Katz, *Fluid Dynamics and Heat Transfer,* McGraw-Hill Book Co., NY (1958), pp. 426–429.
9. Deissler, R. G., NACA TN3145 (1954).
10. Deissler, R. G., and C. S. Eian, NACA TN2629 (1952).
11. Cheremisinoff, N. P., and E. J. Davis, "Heat Transfer to Two-Phase Stratified Gas-Liquid Flow," *AIChE-DIMP,* Report 5, Clarkson College of Technology, Potsdam, NY (1977).

4

RADIATION
HEAT TRANSFER

Planck's Law

Planck posed the concept that radiation is emitted not in a continuous energy state but in discrete quantities (called quanta). The intensity of radiation from a black body is:

$$I_{b,\lambda} = \frac{2c^2 h_0 \lambda^{-5}}{\exp\left(\dfrac{ch_0}{\varkappa \lambda T}\right) - 1} \tag{4-1}$$

where $I_{b,\lambda}$ = radiation of intensity from a black body between wavelengths λ and $\lambda + d\lambda$
- c = speed of light
- h_0 = Planck's constant
- $\varkappa$ = Boltzmann constant
- T = absolute temperature

The total emissive power between wavelengths λ and $\lambda + d\lambda$ is:

$$E_{b,\lambda} = \frac{2\pi c^2 h_0 \lambda^{-5}}{\exp\left(\dfrac{ch_0}{\varkappa \lambda T}\right) - 1} \tag{4-2}$$

Stefan-Boltzmann Law

Planck's law of radiation can be integrated over wavelengths from 0 to ∞ to determine the total emissive power:

$$E_b = \int_0^\infty E_{b,\lambda}d\lambda = \frac{2\pi^5 \varkappa^4 T^4}{15c^2 h_0^3} = \sigma T^4 \qquad (4\text{-}3)$$

where σ is the Stefan-Boltzmann constant (0.1714×10^{-8} Btu/hr-ft²-°R⁴).

Emissivity of Surfaces

Emissivity and absorptivity are the controlling properties in heat transmission via radiation. For black-body radiation Equation 4-3 applies. For actual surfaces:

$$E = \epsilon E_b \qquad (4\text{-}4)$$

where ϵ is the emissivity of a surface. Emissivity varies with wavelengths as well as the direction of emission.

Several generalizations concerning the emissivity of surfaces are:

- Highly polished metallic surfaces usually have very low emissivities
- ϵ depends on surface conditions
- ϵ increases with temperature for all metallic surfaces
- The formation of oxide layers and surface roughness increase ϵ
- ϵ values for nonmetallic surfaces are much higher than for metallic surfaces and shows a decrease as temperature increases.

Absorptivity α of surfaces depends on the quality of the incident radiation. For most materials, in the temperature range of ambient to 2000°F, a good approximation is $\alpha = \epsilon$.

If the incident radiation is derived from a high-temperature source (e.g., solar radiation $\sim$ 10,000°F), emissivity and absorptivity of ordinary surfaces differ greatly. White metal oxides typically have an ϵ-value (and α) of about 0.95 at ordinary temperatures, but α drops sharply to 0.15 if the oxides are exposed to solar radiation. Freshly polished metallic surfaces have an ϵ-value of about 0.05.

A gray surface refers to a surface where the total average absorptivity is independent of the spectral-energy distribution of the incident radiation. This means that ϵ may be used in place of α, even though the temperature of the incident radiation and the receiver are not the same. Examples of good approximations of a gray surface are slate, tar board, and dark linoleum. Table 4-1 lists emissivities of different materials at various temperatures.

Table 4-1
Normal Total Emissivities of Different Surfaces

Surface	t (°F)	Emissivity
Metals		
Aluminum (highly polished, 98.3% pure)	440 ~ 1070	0.039 ~ 0.057
Brass (highly polished)		
73.2% Cu, 26.7% Zn	476 ~ 674	0.028 ~ 0.031
82.9% Cu, 17.0% Zn	530	0.030
Copper		
Polished	242	0.023
Plate heated @ 1110°F	390 ~ 1110	0.57
Molten-state	1970 ~ 2330	0.16 ~ 0.13
Gold	440 ~ 1160	0.018 ~ 0.035
Iron and steel:		
Polished, electrolytic iron	350 ~ 440	0.052 ~ 0.064
Polished iron	800 ~ 1800	0.144 ~ 0.377
Sheet iron	1650 ~ 1900	0.55 ~ 0.60
Cast iron	1620 ~ 1810	0.60 ~ 0.70
Lead (unoxidized)	260 ~ 440	0.057 ~ 0.075
Mercury	32 ~ 212	0.09 ~ 0.12
Nickel (technically pure, polished)	440 ~ 710	0.07 ~ 0.087
Platinum (pure)	440 ~ 1160	0.054 ~ 0.104
Silver (pure)	440 ~ 1160	0.0198 ~ 0.0324
Refractories and miscellaneous materials		
Asbestos	74 ~ 700	0.93 ~ 0.96
Brick, red	70	0.93
Carbon		
Filament	1900 ~ 2560	0.526
Candle soot	206 ~ 520	0.952
Lampblack	100 ~ 700	0.945
Glass	72	0.937
Gypsum	70	0.903
Plaster	50 ~ 190	0.91
Porcelain, glazed	72	0.924
Rubber	75	0.86 ~ 0.95
Water	32 ~ 212	0.95 ~ 0.963

Radiation Between Surfaces

For radiation between the surfaces separated by a nonabsorbing medium, the net loss of energy from a body at room temperature T_1 in black surroundings T_2 is:

$$q_{1,net} = 0.173 \ A_1 \left[\epsilon_1 \left(\frac{T_1}{100} \right)^4 - \alpha_{1,2} \left(\frac{T_2}{100} \right)^4 \right] \tag{4-5}$$

where $q_{1,net}$ is in Btu/hr.
For $\alpha_{1,2} = \epsilon_1$:

$$q_{1,net} = 0.173 \ A_1 \epsilon_1 \left[\left(\frac{T_1}{100} \right)^4 - \left(\frac{T_2}{100} \right)^4 \right] \tag{4-6}$$

The intensity of thermal radiation from a black body, measured at a specified distance and direction from the body, is identical with that which would be emitted from any other black body at the same temperature. The elements of the body's perimeter as viewed from the measuring point must be identical in direction to the corresponding elements of the original black body.

The general equation describing direct radiant-heat exchange between two black surfaces is:

$$q_{net} = (W_{B1} - W_{B2})AF = 0.173 \left[\left(\frac{T_1}{100} \right)^4 - \left(\frac{T_2}{100} \right)^4 \right] AF \tag{4-7}$$

where A = area of one of the surfaces
  F = geometrical factor dependent on the shape and relative orientation of the two surfaces
  W = dq/dA

Values of F as reported by Hottel[1] for various surface arrangements are given in Figure 4-1. F -values are based on the assumptions that ϵ_θ is constant, independent of θ (this is exact for black bodies and a good approximation for most metallic, oxidized, or rough metal surfaces). Figure 4-1B gives F values for adjacent rectangles in perpendicular planes. For opposed parallel rectangles and disks of equal sizes use lines 1 ∿ 4 of Figure 4-1C; for an infinite plane parallel to a system of parallel tubes use lines 1 and 3 of Figure 4-1D.

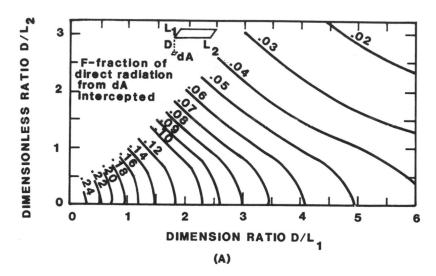

(A)

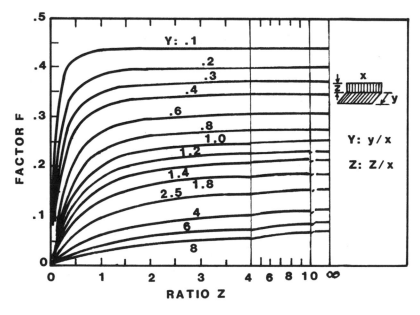

(B)

{figure continued on next page}

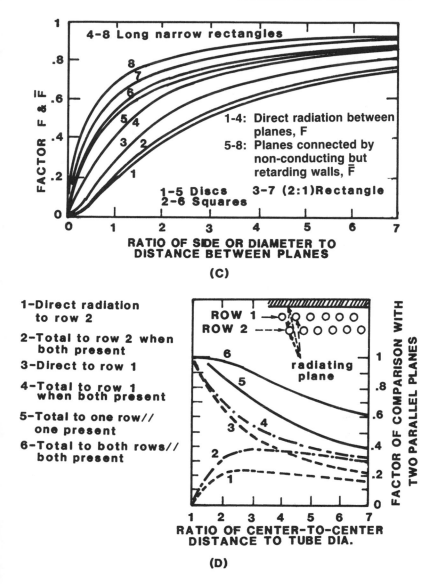

Figure 4-1. (A) Geometrical factor F for direct radiation between element dA and a parallel rectangle, in terms of the dimension ratios; (B) case of radiation between adjacent rectangles in perpendicular planes; (C) parallel planes, directly opposed; (D) radiation between a plane and rows of tubes parallel to the plane.

The net interchange of radiation between two black surfaces is:

$$q = \sigma A_1 F_{1,2}(T_1^4 - T_2^4) \tag{4-8}$$

$$q = \sigma A_2 F_{2,1}(T_1^4 - T_2^4) \tag{4-9}$$

A practical problem is that of radiation to refractory surfaces. An idealized case is an enclosure composed in part of black heat sources and sinks A_1, A_2, ... and refractory surfaces A_R, A_S, ... There is no net radiant heat flux in this case. Unknown refractory surface temperatures can be eliminated from heat balances to give an expression for the net flux from A_1 to A_2:

$$q_{1 \rightleftharpoons 2} = A_1 \bar{F}_{1,2} \sigma(T_1^4 - T_2^4) \equiv A_2 F_{2,1} \sigma(T_1^4 - T_2^4) \tag{4-10}$$

Factor $\bar{F}$ is given for several surface arrangements in Figure 4-1 (lines 5 $\curlywedge$ 8, Figure 4-1C; $F_{2,1}$ is given by line 5 or 6 of Figure 4-1D).

For gray surfaces, the net radiant interchange between A_1 and A_2 (caused by direct radiation, retardation from refractory surfaces, and multiple reflections inside the enclosure).

$$q_{1 \rightleftharpoons 2} = A_1 F_{1,2} \sigma(T_1^4 - T_2^4) \equiv A_2 F_{2,1} \sigma(T_1^4 - T_2^4) \tag{4-11}$$

For two nonrefractory surfaces A_1, A_2, with refractory zones, the F-factor can be evaluated from:

$$F_{1,2} = \cfrac{1}{\cfrac{1}{\bar{F}_{1,2}} + \left(\cfrac{1}{\epsilon_1} - 1\right) + \cfrac{A_1}{A_2}\left(\cfrac{1}{\epsilon_2} - 1\right)} \tag{4-12}$$

Hottel[1] provides sample calculations.

Miscellaneous Data

Detailed discussions of principles of radiation can be found in References 1–3. Tables 4-2 through 4-4 contain useful data. Table 4-2 gives the transmittance of metal foils for monochromatic radiation. Table 4-3 gives transmittance values of electric nonconductors for incident black radiation. Table 4-4 lists geometric mean beam lengths for gas radiation.

Table 4-2
Transmittance of Metal Foils
(Monochromatic Radiation)

Wavelength	(Thickness in microns)*		
λ	Silver	Gold	Platinum
2.5	—	0.0272	0.0352
2.0	—	0.0238	0.0331
1.5	0.0222	0.0244	0.0309
1.0	0.0230	0.0266	0.0284
0.8	0.0236	0.0281	0.0274
0.7	0.0233	0.0311	0.0268

* *Thickness refers to distance to which the intensity of radiation decreases to 1/10 of original value.*
Source: E. Hagen and H. Rubens, Ann. Phys., 8:432 (1902).

Table 4-3
Transmittance of Electric Nonconductors for
Incident Black Radiation (95°C)

Material	Thickness (mm)*
Glass	0.3
Writing paper	0.07
Tracing paper	0.09
Mica	0.25

* *Thickness refers to distance to which the intensity of incident radiation decreases to 1/10 of original value.*
Source: E. Eckert, Forschung Geb. Ingenieurw., 9:251 (1938).

Table 4-4
Geometric Mean Beam Lengths for Gas Radiation[2]

Geometry	Characterizing Dimension	Beam Length (Fraction of Dimension)
Sphere	Diameter	0.667
Infinite Cylinder	Diameter	1
Rectangular Parallelpipeds:		
1:1:1 (cube)	Edge	0.667
1:1:4, radiating to 1 × 4 face	Edge	0.90
radiating to 1 × 1 face	Shortest edge	0.86
radiating to all faces	Shortest edge	0.89
1:∞:∞ (infinite parallel planes)	Distance between planes	2
Space outside infintite bank of tubes with centers on equilaterial triangles	Clearance (= Tube diameter)	3.4

References

1. Hottel, H. C., "Radiant Heat Transmission," in *Heat Transmission*, W. H. McAdams (Ed.), 2nd Ed., McGraw-Hill Book Co., New York (1942), Ch. 3, pp. 45–85.
2. Rosenow, W. M., and J. P. Harnett (Eds.), *Handbook of Heat Transfer*, McGraw-Hill Book Co., New York (1973).
3. Welty, J. R., C. E. Wicks, and R. E. Wilson, *Fundamentals of Momentum, Heat and Mass Transfer*, John Wiley and Sons, Inc., New York (1969).

5

HEAT EXCHANGER CLASSIFICATION

Types of Heat Exchangers

Heat transfer equipment can be specified either by service or type of construction. Various equipment terms are given in the glossary and only principle types are briefly described here. Table 5-1 lists major types of heat exchangers.

The most well-known design is the shell-and-tube heat exchanger. It has the advantages of being inexpensive and easy to clean and available in many sizes, and it can be designed for moderate to high pressure without excessive cost. Figure 5-1 illustrates its design features, which include a bundle of parallel tubes enclosed in a cylindrical casing called a shell.

The basic types of shell-and-tube exchangers are the fixed-tube sheet unit and the partially restrained tube sheet. In the former both tube sheets are fastened to the shell. In this type of construction, differential expansion of the shell and tubes due to different operating metal temperatures or different materials of construction may require the use of an expansion joint or a packed joint. The second type has only one restrained tube sheet located at the channel end. Differential expansion problems are avoided by using a freely riding floating tube sheet or U-tubes at the other end. Also, the tube bundle of this type is removable for maintenance and mechanical cleaning on the shell side.

Shell-and-tube exchangers are generally designed and fabricated to the standards of the Tubular Exchanger Manufacturers Association (TEMA). The TEMA standards list three mechanical standards classes of exchanger construction: R, C, and B.

There are large numbers of applications that do not require this type of construction. These are characterized by low fouling and low corrosivity tendencies. Such units are considered low-maintenance items.

Table 5-1
Summary of Types of Heat Exchangers

Type	Major Characteristics	Application
Shell and tube	Bundle of tubes encased in a cylindrical shell	Always the first type of exchanger to consider
Air cooled heat exchangers	Rectangular tube bundles mounted on frame, with air used as the cooling medium	Economic where cost of cooling water is high
Double pipe	Pipe within a pipe; inner pipe may be finned or plain	For small units
Extended surface	Externally finned tube	Services where the outside tube resistance is appreciably greater than the inside resistance. Also used in debottlenecking existing units
Brazed plate fin	Series of plates separated by corrugated fins	Cryogenic services: all fluids must be clean
Spiral wound	Spirally wound tube coils within a shell	Cryogenic services: fluids must be clean
Scraped surface	Pipe within a pipe, with rotating blades scraping the inside wall of the inner pipe	Crystallization cooling applications
Bayonet tube	Tube element consists of an outer and inner tube	Useful for high temperature difference between shell and tube fluids
Falling film coolers	Vertical units using a thin film of water in tubes	Special cooling applications
Worm coolers	Pipe coils submerged in a box of water	Emergency cooling
Barometric condenser	Direct contact of water and vapor	Where mutual solubilities of water and process fluid permit
Cascade coolers	Cooling water flows over series of tubes	Special cooling applications for very corrosive process fluids
Impervious graphite	Constructed of graphite for corrosion protection	Used in very highly corrosive heat exchange services

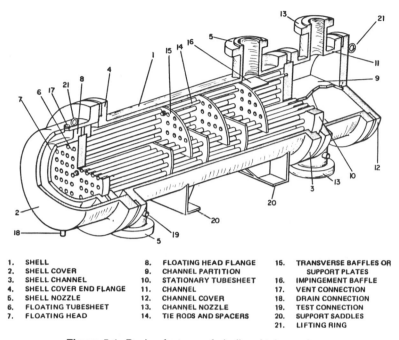

1.	SHELL	8.	FLOATING HEAD FLANGE	15.	TRANSVERSE BAFFLES OR
2.	SHELL COVER	9.	CHANNEL PARTITION		SUPPORT PLATES
3.	SHELL CHANNEL	10.	STATIONARY TUBESHEET	16.	IMPINGEMENT BAFFLE
4.	SHELL COVER END FLANGE	11.	CHANNEL	17.	VENT CONNECTION
5.	SHELL NOZZLE	12.	CHANNEL COVER	18.	DRAIN CONNECTION
6.	FLOATING TUBESHEET	13.	CHANNEL NOZZLE	19.	TEST CONNECTION
7.	FLOATING HEAD	14.	TIE RODS AND SPACERS	20.	SUPPORT SADDLES
				21.	LIFTING RING

Figure 5-1. Design features of shell-and-tube exchangers.

Services falling in this category are water-to-water exchangers, air coolers, and similar non-hydrocarbon applications, as well as some light duty hydrocarbon services such as light ends exchangers, offsite lube oil heaters, and some tank suction heaters. For such services, Class C construction is usually considered. Although units fabricated to either Class R or Class C standards comply with all the requirements of the pertinent codes, (ASME or other national codes), Class C units are designed for maximum economy and may result in a cost saving over Class R.

Air-cooled heat exchangers are another major type composed of one or more fans and one or more heat transfer bundles mounted on a frame. Bundles normally consist of finned tubes. The hot fluid passes through the tubes, which are cooled by air supplied by the fan. The choice of air coolers or condensers over conventional shell-and-tube equipment depends on economics.

Air-cooled heat exchangers should be considered for use in locations requiring cooling towers, where expansion of once-through cooling water systems would be required, or where the nature of cooling causes frequent fouling problems. They are frequently used to remove high level heat, with water cooling used for final "trim" cooling.

These designs require relatively large plot areas. They are frequently mounted over pipe racks and process equipment such as drums and exchangers, and it is therefore important to check the heat losses from surrounding equipment to evaluate whether there is an effect on the air inlet temperature.

Double-pipe exchangers are another class that consists of one or more pipes or tubes inside a pipe shell. These exchangers almost always consist of two straight lengths connected at one end to form a U or "hairpin." Although some double-pipe sections have bare tubes, the majority have longitudinal fins on the outside of the inner tube. These units are readily dismantled for cleaning by removing a cover at the return bend, disassembling both front end closures, and withdrawing the heat transfer element out the rear.

This design provides countercurrent or true cocurrent flow, which may be of particular advantage when very close temperature approaches or very long temperature ranges are needed. They are well suited for high pressure applications, because of their relatively small diameters. Designs incorporate small flanges and thin wall sections, which are advantageous over conventional shell-and-tube equipment. Double-pipe sections have been designed for up to 16,500 kPa gauge on the shell side and up to 103,400 kPa gauge on the tube side. Metal-to-metal ground joints, ring joints, or confined O-rings are used in the front end closures at lower pressures. Commercially available single tube double-pipe sections range from 50-mm through 100-mm nominal pipe size shells with inner tubes varying from 20-mm to 65-mm pipe size.

Designs having multiple tube elements contain up to 64 tubes within the outer pipe shell. The inner tubes, which may be either bare or finned, are available with outside diameters of 15.875 mm to 22.225 mm. Normally only bare tubes are used in sections containing more than 19 tubes. Nominal shell sizes vary from 100 mm to 400 mm pipe.

Extended surface exchangers are composed of tubes with either longitudinal or transverse helical fins. An extended surface is best employed when the heat transfer properties of one fluid result in a high resistance to heat flow and those of the other fluid have a low resistance. The fluid with the high resistance to heat flow contacts the fin surface.

Spiral tube heat exchangers consist of a group of concentric spirally wound coils, which are connected to tube sheets. Designs include countercurrent flow, elimination of differential expansion problems, compactness, and provision for more than two fluids exchanging heat. These units are generally employed in cryogenic applications.

Scraped-surface exchangers consist of a rotating element with a spring-loaded scraper to wipe the heat transfer surface. They are generally used in plants where the process fluid crystallizes or in units where the fluid is extremely fouling or highly viscous.

These units are of double-pipe construction. The inner pipe houses the scrapers and is available in 150-, 200-, and 300-mm nominal pipe sizes. The exterior pipe forms an annular passage for the coolant or refrigerant and is sized as required. Up to ten 300 mm sections or twelve of the smaller individual horizontal sections, connected in series or series/parallel and stacked in two vertical banks on a suitable structure, is the most common arrangement. Such an arrangement is called a "stand."

A bayonet-type exchanger consists of an outer and inner tube. The inner tube serves to supply the fluid to the annulus between the outer and inner tubes, with the heat transfer occurring through the outer tube only. Frequently, the outer tube is an expensive alloy material and the inner tube is carbon steel. These designs are useful when there is an extremely high temperature difference between shell side and tube side fluids, since all parts subject to differential expansion are free to move independently of each other. They are used for change-of-phase service where two-phase flow against gravity is undesirable. These units are sometimes installed in process vessels for heating and cooling purposes. Costs per unit area for these units are relatively high.

Worm coolers consist of pipe coils submerged in a box filled with water. Although worm coolers are simple in construction, they are costly on a unit area basis. Thus they are restricted to special applications, such as a case where emergency cooling is required and there is but one water-supply source. The box contains enough water to cool liquid pump-out in the event of a unit upset and cooling water failure.

A direct contact condenser is a small contacting tower through which water and vapor pass together. The vapor is condensed by direct contact heat exchange with water droplets. A special type of direct contact condenser is a barometric condenser that operates under a vacuum. These units should be used only where coolant and process fluid mutual solubilities are such that no water pollution or product contamination problems are created. Evaluation of process fluid loss in the coolant is an important consideration.

A cascade cooler is composed of a series of tubes mounted horizontally, one above the other. Cooling water from a distributing trough drips over each tube and into a drain. Generally, the hot fluid flows countercurrent to the water. Cascade coolers are employed only where the process fluid is highly corrosive, such as in sulfuric acid cooling.

Impervious graphite heat exchangers are used only in highly corrosive heat exchange service. Typical applications are in isobutylene extraction, dimer and acid concentration plants. The principal construction types are cubic graphite, block type, and shell-and-tube graphite exchangers. Cubic graphite exchangers consist of a center cubic block of impervious graphite that is cross drilled to provide passages for the process and service fluids. Headers are bolted to the sides of the cube to provide for fluid distribution. Also, the cubes can be interconnected to obtain additional surface area. Block-type graphite exchangers consist of an impervious graphite block enclosed in a cylindrical shell. The process fluid (tube side) flows through axial passages in the block, and the service fluid (shell side) flows through cross passages in the block. Shell-and-tube-type graphite exchangers are like other shell-and-tube exchangers except that the tubes, tube sheets, and heads are constructed of impervious graphite.

Shell-and-Tube Exchangers

This subsection provides general information on shell-and-tube heat exchanger layout and flow arrangements. Design details are concerned with several issues—principal ones being the number of required shells, the type and length of tubes, the arrangement of heads, and the tube bundle arrangement.

The total number of shells necessary is largely determined by how far the outlet temperature of the hot fluid is cooled below the outlet temperature of the other fluid (known as the "extent of the temperature cross"). The "cross" determines the value of F_n, the temperature correction factor; this factor must always be equal to or greater than 0.800. (The value of F_n drops slowly between 1.00 and 0.800, but then quickly approaches zero. A value of F_n less than 0.800 cannot be predicted accurately from the usual information used in process designs.) Increasing the number of shells permits increasing the extent of the cross and/or the value of F_n.

The total number of shells also depends on the total surface area since the size of the individual exchanger is usually limited because of handling considerations.

Exchanger tubes are commonly available with either smooth or finned outside surfaces. Selection of the type of surface is based on applicability, availability, and cost.

The conventional shell-and-tube exchanger tubing is the smooth surface type that is readily available in any material used in exchanger manufacture and in a wide range of wall thicknesses. With low-fin tubes, the fins increase the outside area to approximately $2^1/2$ times that of a smooth tube.

Tube length is affected by availability and economics. Tube lengths up to 7.3 m are readily obtainable. Longer tubes (up to 12.2 m for carbon steel and 21.3 m for copper alloys) are available in the United States.

The cost of exchanger surface depends upon the tube length, in that the longer the tube, the smaller the bundle diameter for the same area. The savings result from a decrease in the cost of shell flanges with only a nominal increase in the cost of the longer shell. In the practical range of tube lengths, there is no cost penalty for the longer tubes since length extras are added for steel only over 7.3 m and for copper alloys over 9.1 m.

A disadvantage of longer tubes in units (e.g., condensers) located in a structure is the increased cost of the longer platforms and additional structure required. Longer tube bundles also require greater tube pulling area, thereby possibly increasing the plot area requirements.

Exchanger tubing is supplied on the basis of a nominal diameter and either a minimum or average wall thickness. For exchanger tubing, the nominal tube diameter is the outside tube diameter. The insider diameter varies with the nominal tube wall thickness and wall thickness tolerance. Minimum wall tubing has only a plus tolerance on the wall thickness, resulting in the nominal wall thickness being the minimum thickness. Since average wall tubing has a plus-or-minus tolerance, the actual wall thickness can be greater or less than the nominal thickness. The allowable tolerances vary with the tube material, diameter, and fabrication method.

Tube inserts are short sleeves inserted into the inlet end of a tube. They are used to prevent erosion of the tube itself due to the inlet turbulence when erosive fluids are handled, such as streams containing solids. When it is suspected that the tubes will be subject to erosion by solids in the tube side fluid, tube inserts should be specified. Insert material, length, and wall thickness should be given. Also, inserts are occasionally used in cooling-water service to prevent oxygen attack at the tube ends. Inserts should be cemented in place.

The recommended TEMA head types are shown in Figure 5-2. The *stationary front head* of shell-and-tube exchangers is commonly referred to as the channel. The common TEMA stationary head types and their applications are as follows:

Type A—Features a removable channel with a removable cover plate. It is used with fixed-tube sheet, U-tube, and removable-bundle exchanger designs. This is the most common stationary head type.

Type B—Features a removable channel with an integral cover. It is used with fixed-tube sheet, U-tube, and removable-bundle exchanger design.

Type C—The channel with a removable cover is integral with the tube sheet. There are two varieties of Type C heads available. One is attached to the shell by a flanged joint and is used for U-tube and removable bundles. The other is integral with the shell and is used with fixed-tube sheet designs. The use of Type C heads with U-tube and removable bundles is not recommended since the channel is integral with the tube bundle, which complicates bundle maintenance.

Type D—This is a special high pressure head used when the tube-side design pressure exceeds approximately 6,900 kPa gauge. The channel and tube sheet are integral forged construction. The channel cover is attached by special high pressure bolting.

The TEMA rear head nomenclature defines the exchanger tube bundle type and common arrangements as follows:

Type L—Similar in construction to the Type A stationary head. It is used with fixed-tube sheet exchangers when mechanical cleaning of the tubes is required.

Type M—Similar in construction to the Type B stationary head. It is used with fixed-tube sheet exchangers.

Type N—Similar in construction to the Type C stationary head. It is used with fixed-tube sheet exchangers.

Type P—Called an outside packed floating head. The design features an integral rear channel and tube sheet with a packed joint seal (stuffing box) against the shell. It is not normally used due to the tendency of packed joints to leak. It should not be used with hydrocarbons or toxic fluids on the shell side

Type S—Constructed with a floating tube sheet contained between a split-ring and a tube-sheet cover. The tube sheet assembly is free to move within the shell cover. (The shell cover must be a removable design to allow access to the floating head assembly.)

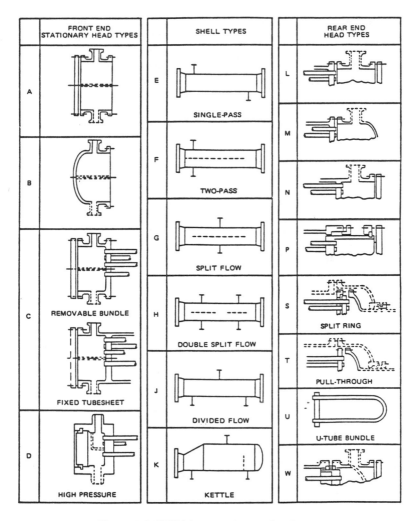

Figure 5-2. TEMA heat exchanger head types.

Type T—Constructed with a floating tube sheet bolted directly to the tube sheet cover. It can be used with either an integral or removable (common) shell cover.

Type U—This head type designates that the tube bundle is constructed of U-tubes.

Type W—A floating head design that utilizes a packed joint to separate the tube-side and shell-side fluids. The packing is compressed against the tube sheet by the shell/rear cover bolted joint. It should *never* be used with hydrocarbons or toxic fluids on either side.

Tube bundles are designated by TEMA rear head nomenclature (see Figure 5-2). Principal types are briefly described below.

Fixed-tube sheet exchangers have both tube sheets attached directly to the shell and are the most economical exchangers for low design pressures. This type of construction should be considered when no shell-side cleaning or inspection is required, or when in-place shell-side chemical cleaning is available or applicable. Differential thermal expansion between tubes and shell limits applicability to moderate temperature differences.

Welded fixed-tube sheet construction cannot be used in some cases because of problems in welding the tube sheets to the shells. Some material combinations that rule out fixed-tube sheets for this reason are carbon steel with aluminum or any of the high copper alloys (TEMA—Rear Head Types L, M, or N).

U-tube exchangers represent the greatest simplicity of design, requiring only one tube sheet and no expansion joint or seals while permitting individual tube differential thermal expansion. U-tube exchangers are the least expensive units for high tube-side design pressures. The tube bundle can be removed from the shell, but replacement of individual tubes (except for ones on the outside of the bundle) is impossible.

Although the U-bend portion of the tube bundle provides heat transfer surface, it is ineffective compared to the straight tube length surface area. Therefore, when the effective surface area for U-tube bundles is calculated, only the surface area of the straight portions of the tubes is included (TEMA—Rear Head Type U).

A pull-through floating head exchanger has a fixed tube sheet at the channel end and a floating tube sheet with a separate cover at the rear end. The bundle can be easily removed from the shell by disassembling only the front cover. The floating head flange and bolt design require a relatively large clearance between the bundle and shell, particularly as the design pressures increase. Because of this clearance, the pull-through bundle has fewer tubes per given shell size than other types of construction do. The bundle-to-shell clearance, which decreases shell-side heat transfer capability, should be blocked by sealing strips or dummy tubes to reduce shell-size fluid bypassing. Mechanical cleaning of both the shell and tube sides is possible (TEMA—Rear Head Type T).

A split-ring floating head exchanger has a fixed-tube sheet at the channel end and a floating tube sheet that is sandwiched between a split-ring and a separate cover. The floating head assembly moves inside a shell cover of a larger diameter than that of the shell. Mechanical cleaning of both the shell and tube is possible (TEMA—Rear Head Type S).

There are two variations of *outside packed floating head designs*, the lantern ring type and the stuffing box type. In the lantern ring design, the floating head slides against a lantern ring packing, which is compressed between the shell flange and the shell cover. The stuffing box design is similar to the lantern ring type, except that the seal is against an extension of the floating tube sheet and the tube sheet cover is attached to the tube sheet extension by means of a split-ring. (TEMA— Rear Head Types P or W).

Tube Arrangements and Baffles

The following are some general notes on tube layout and baffle arrangements for shell-and-tube exchangers. There are four types of tube layouts with respect to the shell-side crossflow direction between baffle tips: square (90°), rotated square (45°), triangular (30°), and rotated triangular (60°). The four types are shown in Figure 5-3.

Use of triangular layout (30°) is preferred (except in some reboilers). An exchanger with triangular layout costs less per square meter and transfers more heat per square meter than one with a square or rotated square layout. For this reason, triangular layout is preferred where applicable.

Rotated square layouts are preferable for laminar flow, because of a higher heat transfer coefficient caused by induced turbulence. In turbulent flow, especially for pressure-drop limited cases, square layout is preferred since the heat transfer coefficient is equivalent to that of rotated square layout while the pressure drop is somewhat less.

Tube layout for removable bundles may be either square (90°), rotated square (45°), or triangular (30°). Nonremovable bundles (fixed-tube sheet exchangers) are *always* triangular (30°) layout.

The *tube pitch* (PT) is defined as the center-to-center distance between adjacent tubes (see Figure 5-3). Common pitches used are given in Table 5-2.

The column "Heaviest Recommended Wall" is based on the maximum allowable tube sheet distortion resulting from rolling the indicated tube into a tube sheet having the minimum permissible ligament width at the listed pitch. The ligament is that portion of the tube sheet between two adjacent tube holes.

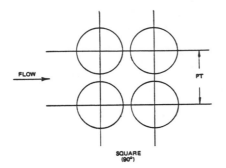

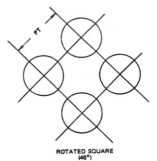

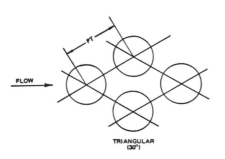

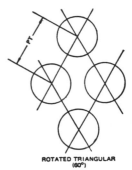

Figure 5-3. Tube layouts.

Table 5-2
Common Tube Pitch Values

Tube Size	Triangular (mm)	Square (mm)	Heaviest Recommended Wall (mm)
19.05 mm O.D.	23.81	—	2.41
19.05 mm O.D.	25.40	25.40	2.77
25.4 mm O.D.	31.75	31.75	3.40
38.1 mm O.D.	47.63	47.63	4.19
> 38.1 mm	Use 1.25 times the outside diameter		

Tubes are supported by baffles that restrain tube vibration from fluid impingement and channel fluid flow on the shell side. Two types of baffles are generally used: segmental and double segmental. Types are illustrated in Figure 5-4.

The baffle cut is the portion of the baffle "cut" away to provide for fluid flow past the chord of the baffle. For segmental baffles, this is the ratio of the chord height to shell diameter in percent. Segmental baffle cuts are usually about 25%, although the maximum practical cut for tube support is approximately 48%.

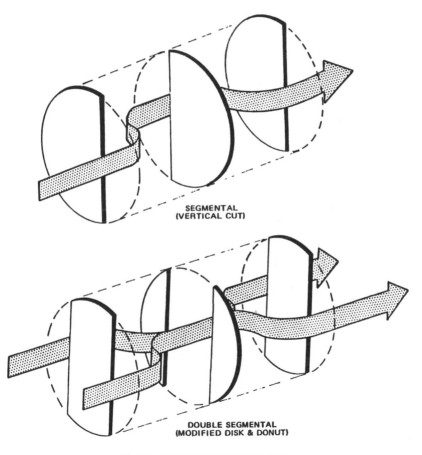

SEGMENTAL
(VERTICAL CUT)

DOUBLE SEGMENTAL
(MODIFIED DISK & DONUT)

Figure 5-4. Types of shell baffles.

Double segmental baffle cut is expressed as the ratio of window area to exchanger cross sectional area in percent. Normally the window areas for the single central baffle and the area of the central hole in the double baffle are equal and are 40% of the exchanger cross-sectional area. This allows a baffle overlap of approximately 10% of the exchanger cross-sectional area on each side of the exchanger. However, there must be enough overlap so that at least one row of tubes is supported by adjacent segments.

Baffle pitch is defined as the longitudinal spacing between baffles. The maximum baffle pitch is a function of tube size and, for no change of phase flow, of shell diameter. If there is no change of phase in the shell-side fluid, the baffle pitch should not exceed the shell inside diameter. Otherwise, the fluid would tend to flow parallel with the tubes, rather than perpendicular to them, resulting in a poorer heat transfer coefficient.

Impingement baffles are required on shell-side inlet nozzles to protect the bundle against impingement by the incoming fluid when the fluid:

1. Is condensing
2. Is a liquid vapor mixture
3. Contains abrasive material
4. Is entering at high velocity

In addition, TEMA requires bundle impingement protection when nozzle values of ρu^2 (fluid density, kg/m^3, times velocity squared m^2/s^2) exceed:

1. 2230 $kg/m\text{-}s^2$ for noncorrosive, nonabrasive, single-phase fluids
2. 744 $kg/m\text{-}s^2$ for all other liquids

Also, the minimum bundle entrance area should equal or exceed the inlet nozzle area and should not produce a value of ρu^2 greater than 5,950 $kg/m\text{-}s^2$, per TEMA. Impingement baffles can be either flat or curved. In order to maintain a maximum tube count, the impingement plate is sometimes located in a conical nozzle opening or in a dome cap above the shell. The impingement plate material should be at least as good as that of the tubes.

Shell Configurations

The following notes summarize design features of shells for shell-and-tube heat exchangers.

The *single-pass shell* is the most common shell construction used. The shell-side inlet and outlet nozzles are located at opposite ends of the shell. The nozzles can be placed on opposite or adjacent sides of the shell, depending on the number and type of baffles used. A typical one-shell pass exchanger with horizontal segmental baffles is illustrated in Figure 5-5A (TEMA E).

A *two-pass shell* uses a longitudinal baffle to direct the shell-side flow. An exchanger with two shell passes is shown in Figure 5-5B. Note that both the shell inlet and outlet nozzles are adjacent to the stationary tube sheets. A shell-side temperature range exceeding 195°C should be avoided, since greater temperature ranges result in excessive heat leakage through the baffle, as well as thermal stresses in the baffle, shell, and tube sheet.

The longitudinal baffle can be either welded or removable. Since there are severe design and cost penalties associated with the use of welded baffles in floating head exchangers, this type of design should be used only with fixed-tube sheet units that do not require expansion joints. Removable longitudinal baffles require the use of flexible light gauge sealing strips or a packing device between the baffle and the shell, to reduce fluid leakage from one side to the other (TEMA F).

A *divided flow shell* has a central inlet nozzle and two outlet nozzles, or vice-versa. A divided flow exchanger is illustrated in Figure 5-5C. This type is generally used to reduce pressure drop in a condensing service. In minimizing pressure drop the shell fits in as follows:

- E shell with segmental baffles
- E shell with double segmental baffles
- J shell with segmental baffles
- J shell with double segmental baffles
- E shells in parallel with segmental baffles
- E shells in parallel with double segmental baffles
- J shells in parallel with segmental baffles
- J shells in parallel with double segmental baffles

Generally, for most designs, double segmental baffles are used with J shells.

Double segmental baffles in a divided-flow exchanger normally have a vertical cut. This baffle arrangement also requires that there be an odd number of total baffles, but there must also be an odd number of baffles in each end of the shell. The center baffle for this arrangement is normally similar to the center baffle used with segmental cut. The baffles on each side of the central baffle and the last baffle toward the ends of the shell have solid centers with cutaway edges.

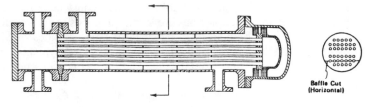

SINGLE PASS SHELL

(A)

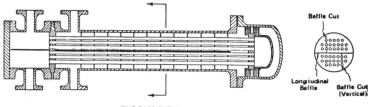

TWO PASS SHELL

(B)

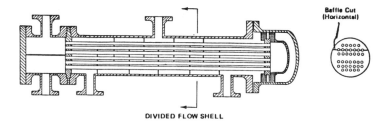

DIVIDED FLOW SHELL

(C)

Figure 5-5. (A) Single pass shell; (B) two-pass shell; (C) divided flow shell.

The choice of whether to stack shells depends on maintenance considerations, as well as on the amount of plot area available. Stacking shells requires less area and frequently less piping. Normally, shells are not stacked more than two high. However, stacked heat exchangers are more costly to maintain, because of accessibility.

If sufficient plot area is available, the following guidelines apply:

1. If the fluids are known to be clean and noncorrosive, the shells should usually be stacked.
2. If the fluids are moderate clean or slightly corrosive, the shells may be stacked.
3. If the fluids are very dirty or corrosive, the shells should not be stacked, to allow for ease of maintenance.

Miscellaneous Data

Design-related data are given in Tables 5-3 through 5-5. Table 5-2 provides typical tube dimensions and tube surface areas per unit length. Table 5-3 gives thermal conductivities of materials commonly used for exchanger construction. Additional data can be found in Chapter 1. Table 5-4 gives recommended maximum number of tube passes as a function of tube size.

Table 5-3
Tube Dimensions and Surface Areas Per Unit Length

d_o = O.D. of Tubing (mm)	δ = Wall Thickness (mm)	d_i = I.D. of Tubing (mm)	Internal Area (mm^2)	External Surface Per Ft Length (sq ft/ft)
19.05	2.77	13.51	143.8	0.0598
19.05	2.11	14.83	172.9	0.0598
19.05	1.65	15.75	194.8	0.0598
19.05	1.24	16.56	215.5	0.0598
25.40	3.40	18.59	271.6	0.0798
25.40	2.77	19.86	309.0	0.0798
25.40	2.11	21.18	352.3	0.0798
25.40	1.65	22.10	383.2	0.0798
38.10	3.40	31.29	769.0	0.1197
38.10	2.77	32.56	832.9	0.1197
38.10	2.11	33.88	901.3	0.1197

Table 5-4
Thermal Conductivities of Materials of Construction

Material	Composition	Thermal Conductivity, k, Wt/m-°C
Admiralty	71 Cu-28 Zn-1 Sn	111
Type 316 stainless steel	17 Cr-12 Ni-2 Mo	16
Type 304 stainless steel	18 Cr-8 Ni	16
Brass	70 Cu-30 Zn	99
Red brass	85 Cu-15 Zn	159
Aluminum brass	76 Cu-22 Zn-2 Al	100
Cupro-nickel	90 Cu-10 Ni	71
Cupro-nickel	70 Cu-30 Ni	29
Monel	67 Ni-30 Cu-1.4 Fe	26
Inconel		19
Aluminum		202
Carbon steel		45
Carbon-moly	0.5 Mo	43
Copper		386
Lead		35
Nickel		62
Titanium		19
Chrome-moly steel	1 Cr-0.5 Mo	42
	2¼ Cr-0.5 Mo	38
	5 Cr-0.5 Mo	35
	12 Cr-1 Mo	28

Table 5-5
Maximum Number of Tube Passes

Shell I.D. (mm)	Recommended Maximum Number of Tube Passes
<250	4
250 – <510	6
510 – <760	8
760 – <1,020	10
1,020 – <1,270	12
1,270 – <1,520	14

References

1. API Standard 661, Air-Cooled Heat Exchangers for General Refinery Services.
2. Standards of Tubular Exchanger Manufacturer's Association (TEMA); BP6-1-1; BP6-1-2, BP6-2-1, BP6-3-1 and BP6-4-1.

6

HEAT EXCHANGER DESIGN NOTES

General Notes

There are two basic design problems encountered with heat exchangers.

1. Rating problems, in which the heat exchanger type, size, and surface geometry and process conditions such as flow rate, inlet fluid conditions, and fouling factors are specified. The analysis centers around calculation of the heat transfer rate and pressure drop of each stream.
2. Sizing problems, in which the heat exchanger size required for specified process requirements (e.g., flow rate, entering fluid conditions and allowable pressure drops) is evaluated.

The latter class of problems is more complex than the rating problem since the heat exchanger type and surface geometry must be selected before the thermal analysis can be made. A preliminary specification covers the heat exchanger materials, the heat exchanger type and flow arrangement, and the heat transfer surface geometries.

Specific design specifications to be considered are as follows:

- Process requirements
  1. Fluid compositions/inlet flow conditions
  2. Heavy duty or required exit temperatures
  3. Allowable pressure drops
- Operating and maintenance considerations
  1. Fouling potential and cleaning method
  2. Failure due to corrosion, thermal stress, vibration, etc.
  3. Repair of leaks
  4. Part load operating characteristics

- Size and weight restrictions
  1. Frontal area/length or height requirements
  2. Possible weight restrictions

Design selection considerations consist of the following:

- Exchanger Materials of Construction
  1. Fluid temperatures and pressures
  2. Corrosive properties of fluid-material combinations
- Heat exchanger types
  1. Design pressures and fluid temperatures
  2. Corrosion, stress, vibration and freezing considerations
  3. Capital equipment investment/operating and maintenance costs
- Heat transfer surface geometries
  1. Thermal resistance ratio of fluids
  2. Potential use of augmented surfaces
  3. Potential for fouling and cleaning options
  4. Unit cost of heat transfer surface

Once the exchanger type has been specified, the thermal and hydraulic analyses are needed. This involves evaluation of several possibilities to ascertain the "optimum" design. The optimum design requires establishing which variables (the objective function) are to be optimized, e.g., first cost, operating cost, or alternatively "life-cycle costs," and size dimensions. This process involves a parametric analysis to determine how the specified objective function is influenced by trade-offs among the possible design variables. References 1 and 2 provide detailed discussions.

Thermal Analysis of Exchangers

The heat exchanged between the two fluids depends on the following:

Thermodynamic specifications—Defined by the hot and cold fluid flow rates and their entering and exit temperatures:

$$q = C_H(T_{H1} - T_{H2}) = C_C(T_{C2} - T_{C1}) \tag{6-1}$$

where $C_H = W_H c_{pH}$, the "capacity rate" of the hot fluid.

The heat transfer rate equation:

$$dq = UdA(T_H - T_c) \tag{6-2}$$

The heat rate depends on the overall heat transfer coefficient between the hot and cold fluids. The heat exchanger surface area required to satisfy the thermodynamic specifications of Equation 6-1 is obtained by integrating the heat transfer rate equation over the length (or area) of the exchanger. For constant U:

$$q = U \int_A (T_H - T_C) dA \tag{6-3}$$

The "effective mean temperature difference," averaged over the total surface area is:

$$\Delta T_M = \frac{1}{A} \int_A (T_H - T_C) dA \tag{6-4}$$

Combining these expressions gives:

$$q = UA\Delta T_m \tag{6-5}$$

The product of the terms UA—Defined as the "overall heat conductance" of the exchanger. The inverse 1/UA is the overall thermal resistance to heat transfer between the two fluids and is composed of individual thermal resistances.

$$\frac{1}{UA} = \frac{1}{(\eta_0 nA)_H} + \left(\frac{R_f}{A}\right)_H + \frac{t}{hA_w} + \frac{1}{(\eta_0 hA)_C} + \left(\frac{R_f}{A}\right)_C \tag{6-6}$$

Subscripts H and C refer to the hot and cold streams. The first and fourth terms in Equation 6-6 account for the convective resistance between the flowing fluids and the pipe wall. The third term is the conduction resistance of the solid wall that separates the streams. The second and fifth terms are the fouling resistance on the hot and cold heat transfer surfaces. The overall coefficient, U, is defined either in terms of the surface area of the hot or cold surface areas:

$$UA = U_H A_H = U_c A_c \tag{6-7}$$

The effective temperature difference ΔT_m is a function of the exchanger's flow geometry, e.g., counterflow and cross-flow, and the degree of fluid mixing within each flow stream. For counterflow and parallel flow geometries:

Counterflow:

$$\Delta T_m = \frac{(T_{H1} - T_{C2}) - (T_{H2} - T_{C1})}{\ln \dfrac{T_{H1} - T_{C2}}{T_{H2} - T_{C1}}} \tag{6-8}$$

Parallel flow:

$$\Delta T_m = \frac{(T_{H1} - T_{C1}) - (T_{H2} - T_{C2})}{\ln \dfrac{T_{H1} - T_{C1}}{T_{H2} - T_{C2}}} \tag{6-9}$$

Denoting ΔT_I and ΔT_{II} as the difference between the fluid temperatures at each end of the exchanger, the logarithmic mean temperature difference is used as the effective temperature difference ΔT_m for the counterflow and parallel flow arrangements.

$$\text{LMTD} = \frac{\Delta T_I - \Delta T_{II}}{\ln (\Delta T_I / \Delta T_{II})} \tag{6-10}$$

For more complex geometries such as crossflow or multipass flow configurations, integration of Equation 6-4 provides more complex expressions for ΔT_m. Alternate techniques for calculating the heat transfer rate between the two fluids are the UA-LMTD and the effectiveness-NTU methods.

The UA-LMTD method—The basis is Equation 6-5 and a correction factor for the log-mean temperature difference is:

$$F = \frac{\Delta T_m}{\text{LMTD}} \tag{6-11}$$

The heat transfer rate is:

$$q = F(UA)(\text{LMTD}) \tag{6-12}$$

The calculation steps in this procedure are as follows:

Step 1: Compute the overall heat transfer coefficient (from Equation 6-6.
Step 2: Compute the LMTD for a counterflow arrangement.
Step 3: Evaluate the geometry correction factor, F.

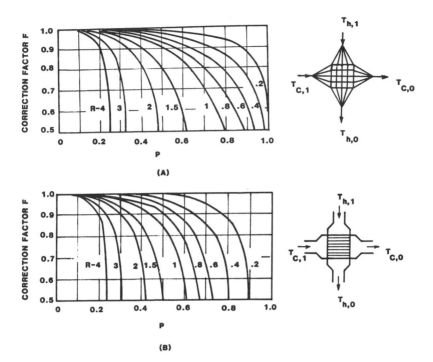

Figure 6-1. Plots of correction factor F: (A) crossflow heat exchanger with both fluids unmixed: $P = (T_{c,o} - T_{c,1}) / (T_{h,1} - T_{c,1})$; $R = (T_{h,1} - T_{h,o}) / (T_{c,o} - T_{c,1})$; (B) crossflow heat exchanger with one fluid mixed and one fluid unmixed. *Note:* When hot fluid flows in tubes, interchange subscripts c and h.

For counterflow and parallel-flow arrangement the value of F is unity. Figure 6-1 shows a correlation for F in terms of variables P and R for two flow geometries. Figure 6-1A is for both fluids unmixed and Figure 6-B is for one fluid mixed and the other unmixed. Unmixed means that the fluid stream flows through the exchanger in separate tubes (i.e., no mixing between neighboring flow tubes). Dimensionless variables P and R are defined in the figure.

Graphs of the F factor are given in References 3 and 4 for other flow arrangements. Multipass exchangers have a combination of parallel and counter flow in alternate passes; hence, F depends on the "pass arrangement," i.e., whether fluid mixing occurs within a pass, and the conditions of mixing between passes.

Effectiveness-NTU method—In this method, the heat exchanger effectiveness is defined as:

$$\epsilon = \frac{\text{actual heat transfer}}{\text{maximum possible heat transfer}} = \frac{q_{actual}}{q_{max}} \qquad (6\text{-}13)$$

Equation 6-1 defines q_{actual}. The maximum possible heat transfer will occur in a counterflow heat exchanger of infinite area if one fluid undergoes a temperature change equal to the maximum temperature difference available.

$$\Delta T_{max} = T_{H1} - T_{C1} \qquad (6\text{-}14)$$

q_{max} is based on the fluid having the smaller capacity rate, C_{min}.

$$q_{max} = C_{min}\Delta T_{max} = C_{min}(T_{H1} - T_{C1}) \qquad (6\text{-}15)$$

and

$$q_{actual} = C_{min}(T_{H1} - T_{C1}) \qquad (6\text{-}16)$$

For a given flow geometry, the heat exchange effectiveness (i.e., number of heat transfer units, NTU) is a function of two dimensionless quantities, $UA/C_{min} \equiv C_{min}/C_{max}$ and R. The effectiveness equation is obtained by algebraic manipulation of the equations developed in calculation of the LMTD. The development of the effectiveness equations for parallel and counterflow are given by Holman.[4] The effectiveness for a counterflow heat exchanger is shown in Figure 6-2, and is given by

$$\epsilon = \frac{1 - e^{-NTU(1 - R)}}{1 - Re^{-NTU(1 - R)}} \qquad (6\text{-}17)$$

The NTU is related to the heat exchanger's physical size. The capacity rate ratio, R, is equal to zero for an evaporator or condenser if the fluid remains at a constant temperature during the phase change. If R = 0, the ϵ-NTU relation is independent of the heat exchanger flow geometry. Figure 6-2 shows that the required NTU increases for increasing values of effectiveness, and asymptotically approaches infinity as the effectiveness approaches its thermodynamic limit ($\epsilon = 1$).

Table 6-1 lists the effectiveness equations for several flow geometries. Equations are listed in order of decreasing effectiveness for con-

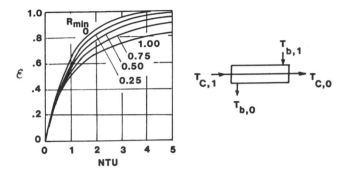

Figure 6-2. Heat exchanger effectiveness for counterflow.

stant R and NTU. The right hand column shows the effectiveness obtainable for R = 1 and NTU = 4.

The NTU method allows physical interpretation of the thermodynamic performance of the exchanger not provided by the LMTD method. Also, the ϵ-NTU relations are readily available in algebraic form; compatible for digital computer calculations.

Fin Efficiency Estimation

Fins are extended surfaces that are applied to the fluid stream having the dominant thermal resistance. They provide reduced thermal resistance through increased surface area. The heat transfer coefficient on the extended surfaces may be either higher or lower than that which would occur on the unfinned surface. Low radial fins on horizontal tubes with condensation provides both an area increase and increased heat transfer coefficient. In contrast, fins in single-phase forced convection can cause slight reductions of the coefficient, depending on the fin spacing. The heat conductance of a finned surface (hA) must be multiplied by a fin efficiency factor to account for the temperature gradient in the fin. The fin efficiency, η_f, is defined as the ratio of the actual heat transfer from the fin to that which would occur if the entire fin were at its base temperature.

$$\eta_f = \frac{q}{hA_f(T_0 - T_\infty)} \tag{6-18}$$

Table 6-1
Equations for Heat Exchanger Effectiveness

Pass Arrangement	Effectiveness	ϵ for R = 1, N = 4
Counterflow	$$\frac{1 - \exp\left[-N(1 - R)\right]}{1 - R\exp\left[-N(1 - R)\right]}$$	.80
Cross-counter flow (4 pass, fluids mixed between passes)	$$\frac{[(1 - \epsilon_p R)/(1 - \epsilon_p)]^4 - 1}{[(1 - \epsilon_p R)/(1 - \epsilon_p)]^4 - R}$$	.785
2-4 parallel-counterflow (Shell mixed)	none given	.72
Cross flow (Both fluids unmixed)	$$1 - \exp\left\{\frac{1}{R\eta}[\exp(-RN\eta)-1]\right\}$$ $$\eta = N^{-.22}$$	.73
Cross flow (One fluid unmixed)	$$\frac{1}{R}\{1 - \exp[R(1 - e^{-N})]\}$$	.625
1-2 Parallel-Counterflow	$$2\left[(1 + R) + (1 + R^2)^{1/2}\frac{(1 + e^{-\Gamma})}{(1 - e^{-\Gamma})}\right]^{-1}$$ $$\Gamma = N(1 + R^2)^{1/2}$$	.582
Cross flow (Both fluids mixed)	$$\left[\frac{1}{1 - e^{-N}} + \frac{R}{1 - e^{-RN}} - \frac{1}{N}\right]^{-1}$$	.559
Parallel flow	$$\frac{1 - \exp[-N(1 + R)]}{1 + R}$$	.50

$N \equiv NTU \equiv UA/C_{min}$
$R \equiv C_{min}/C_{max}$

Fin efficiency depends on the fin's cross-sectional shape, length, and the geometry of the base surface. Figure 6-3 gives the efficiency of plain and circular fins of uniform cross section; applicable for an adiabatic fin tip boundary condition. Fins which have convection from their tips may be determined from Figure 6-3 providing the fin length is modified to the value $(L_c = L + t/2)$.[5]

A finned surface heat exchanger consists of the secondary finned sur-face and the primary surface to which the fins are attached. A second term called the *total surface efficiency* is defined to account for the effi-ciency of the composite structure consisting of the fins and the base sur-face. The total surface efficiency is:

$$\eta = 1 + (1 - \eta_f)\frac{A_f}{A} \tag{6-19}$$

This definition assumes the heat transfer coefficient on the finned sur-face is the same as on the base surface.

Correlations for Heat Transfer Coefficients

Augumented heat transfer surface performance is rated relative to the performance of the smooth or nonaugmented surface. A summary of design equations for the heat transfer of smooth surfaces is given in Ta-bles 6-2 through 6-4.

Table 6-2 lists equations for free convection from horizontal, verti-cal, and inclined plates, enclosures, and three-dimensional bodies. Buoyancy is the driving force in free convection, hence, the equations for upward-facing hot plates are applicable to downward facing cold plates. Different relations are required for constant temperature and constant heat flux boundary conditions.

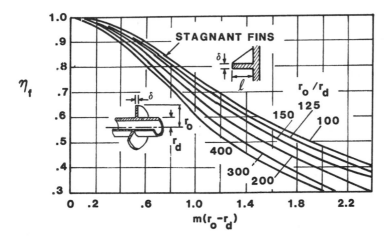

Figure 6-3. Efficiency of circular fins of rectangular cross section.

Table 6-2
Equations for Free Convection

Configuration	Rayleigh Number Range	Coefficients/Correlation
Vertical plates		
T_w = const.	$10^4 < Ra_L < 10^9$	C = .59, n = .25, L_c = L
	$10^9 < Ra < 10^{12}$	C = .10, n = $1/3$ L_c = L
q_w = const.	$10^5 < Ra^* < 10^{11}$	C = .6, n = .2 L_c = L
	$10^{13} < Ra^* < 10^{16}$	C = .57, n = .22 L_c = L
Horizontal plates		
T_w = const., plate facing up	$2.6 \times 10^4 < Ra < 10^7$	C = .54, n = $1/4$ L_c = A/P P = perimeter; A = surface area
	$10^7 < Ra < 3 \times 10^{10}$	C = .15, n = $1/3$ L_c = A/P
T_w = const., plate facing down	$3 \times 10^5 < Ra < 3 \times 10^{10}$	C = .27, n = $1/4$ L_c = A/P
Enclosed spaces	$Nu_\delta = cRa_\delta^n (L/\delta)^m$	
Vertical space T_w = const.	$6,000 < Ra_\delta < 20,000$	C = .197, n = $1/4$, m = $-1/9$
	$2 \times 10^5 < Ra_\delta < 1.1 \times 10^7$	C = .073, n = $1/3$, m = $-1/9$
Inclined space T_w = const.	$0 < \theta < 120°$	$Nu_\theta / Nu_{vert} = (\sin \theta)^{1/4}$
Horizontal space T_w = const.	$6 \times 10^6 < Ra_\delta < 10^8$	$Nu = .104\, Ra_\delta^{.305}\, Pr^{.084}$
Three-dimensional bodies (T_w = const.)		
Horizontal cylinder	$10^4 < Ra < 10^9$	C = .53, n = $1/4$, L_c = D
	$10^9 < Ra < 10^{12}$	C = .13, n = $1/5$, L_c = D
Long vertical cylinder	Same as vertical plate	L_c = L
Rectangular block	$10^4 < Ra < 10^9$	C = .55, n = $1/4$, $L_c = L_h L_v / (L_h + L_v)$ L_h = longer of the two horizontal dimensions, L_v = vertical dimension
Short vertical cylinder	Use rectangular block equation	

$Nu_T = Cra^n$; $Nu_H = C(Ra^*)^n$; $Ra = g\beta L_c^3 \Delta T / v\alpha$; $Ra^* = (g\beta L_c^3 / v\alpha) q_w L_c / k$

Table 6-3
Equations for Forced Convection

	Friction Factor	Nusselt Number
Laminar flow in ducts		
1. Circular tube (F.D.)	$f = 16/Re_D$	$Nu_T = 3.66$ $Nu_H = 4.36$
2. Noncircular ducts (F.D.)	See Table 6-4	
3. Entrance length	See Figure 6-4	
Turbulent flow in ducts		
1. Circular tube (recommended equation)	$f = (1.82 \log_{10} Re - 1.64)^{-2}$	$Nu = (f/2) \, Re \, Pr/$ $[1.07 + 12.7$ $\sqrt{f/2} \, (Pr^{2/3} - 1)]$
2. Circular tube (simple, approximate equation)	$f = .079/Re^{.25}$ $Re < 50{,}000$ $f = .046/Re^{.2}$ $Re > 50{,}000$	$Nu = .023 \, Re^{.8} Pr^n$ $n = 0.4$ (heating) Same as above $n = 0.3$ (cooling)
3. Noncircular tube	Use tube hydraulic diameter for Nu, Re	
Tube banks in crossflow		
1. Inline banks	$f = .025 + \dfrac{.1175(DU_{max}/\nu)^{-.16}}{(S_t/D - 1)^{1.08}}$	$Nu = C(DU_{max}/\nu)^n Pr^{1/3}$
2. Staggered banks	$f = \dfrac{[.044 + .08(S_1/D)](DU_{max})^{.15}/\nu}{(S_t/D - 1)^{(.43 + 1.13D/S_t)}}$	Same as above
3. Staggered and inline banks	$f \sim$ Zukauskas graphical correlation Use $Re = DU_o/\nu$ $\Delta p = f \cdot \dfrac{1}{2} \rho U^{-2} . N$	Zukauskas graphical and empirical curve fits for limited Re ranges $Re = DU_o/\nu$

F.D. = Fully developed

Table 6-4
Fully Developed Laminar Flow Solutions

Geometry	Nu_H	Nu_T	fRe	$K(\infty)$	j_H/f	j_T/f	L_{hy}^+
$2b \rule{2em}{0.4pt}$ $\frac{2b}{2a} = \frac{1}{8}$ $2a$	6.490	5.597	20.585	.879	.355	.306	.0094
$2b \rule{2em}{0.4pt}$ $\frac{2b}{2a} = \frac{1}{6}$ $2a$	6.049	5.137	19.702	.945	.346	.294	.0110
$2b \rule{2em}{0.4pt}$ $\frac{2b}{2a} = \frac{1}{4}$ $2a$	5.331	4.439	18.233	1.076	.329	.274	.0147
$\bigcirc$	4.364	3.657	16.00	1.24	.307	.258	.038
$2b \square$ $\frac{2b}{2a} = \frac{1}{2}$ $2a$	4.123	3.391	15.548	1.383	.299	.245	.0255
$2b \square$ $\frac{2b}{2a} = 1$ $2a$	3.608	3.091	14.227	1.552	.286	.236	.0324
$2b \triangle$ $\frac{2b}{2a} = \frac{\sqrt{3}}{2}$ $2a$	3.111	2.47	13.333	1.818	.263	.209	.0398
$2b \triangle$ $\frac{2b}{2a} = \frac{\sqrt{3}}{2}$ $2a$	3.014	2.39	12.630	1.739	.269	.214	.0408
$2b \triangle$ $\frac{2b}{2a} = 2$ $2a$	2.88	2.22	13.026	1.991	.249	.192	.0443
$\triangle$ $\frac{2b}{2a} = .25$	2.60	1.99	12.622	2.236	.232	.178	.0515

$j/f \equiv \dfrac{NuPr^{-1/3}}{fRe}$; j_H and j_T for $Pr = 0.7$

$H \sim$ *heat flux boundary condition*
$T \sim$ *constant temperature boundary condition*
$K(\infty) \sim$ *entrance factor*

Table 6-3 gives both friction factor and Nusselt number correlations for turbulent duct flow equations that also apply to noncircular ducts and axial flow in rod bundles by replacing the tube size with the hydraulic diameter. The hydraulic diameter does not apply to laminar flow, for which the shape of the channel is important. Table 6-4 gives friction factors and Nusselt numbers for noncircular ducts with fully-developed laminar flow. High-aspect-ratio rectangular channels provide higher heat transfer coefficients and heat transfer per unit of pressure drop (j/f) than circular tubes and triangular channels.

Table 6-5 gives correlations for film condensation on vertical and inclined plates, horizontal tubes, and for a vertical row of horizontal tubes. The relation of $h_n/h_1 = N^{-.25}$ is not applicable to most practical geometries. In tube banks, the effects of film turbulence, splashing or vapor shear may reduce the row effect exponent to values on the order of 1/6 to 1/10.

For augmented heat transfer surfaces, the Reynolds analogy applies ($Nu \propto Pr^{1/3}$), expressed by Colburn[6] as:

$$StPr^{2/3} \equiv j = f/2 \tag{6-20}$$

Equation 6-20 applies to laminar and turbulent flow over flat plates, turbulent flow in smooth ducts or rod bundles, and noncircular channels (using the hydraulic diameter).

The equation $2j/f = 1$ establishes a relation between friction due to surface shear and heat transfer. The analogy does not account for flow separations. With flow across a tube bank or over a rough surface, the total drag force consists of surface shear and form drag. Form drag contributes little to heat transfer.

Augumented heat transfer surfaces produce increased heat transfer coefficients; $j/j_s > 1$, where subscript "s" refers to a smooth surface. For comparison at equal Reynolds numbers, Equation 6-20 implies that the augmented surface must have a friction increase at least as large as the j/j_s increase. Augmented surfaces, in general, show $(j/j_s)/(f/f_s) < 1$.

The relative j/f ratio of augmented-to-smooth surfaces is referred to as the "efficiency index,"

$$\eta_e \equiv \frac{j/j_s}{f/f_s} \tag{6-21}$$

η_e is useful in evaluating the quality of the augmentation. High values for j/j_s with η_e as close as possible to the value 1.0 are desirable. Smaller values of η_e cause increased friction penalty to establish a given augmentation level, j/j_s. In some cases (turbulent flow of high-Prandtl-number liquids over roughness) it is possible to obtain $\eta_e > 1$. In this case the Prandtl-number dependency of the rough surface flow is greater than that of smooth surfaces. For $Pr < 1$, augmented surfaces do not show η_e values greater than unity.

$2j/f$ does not apply to laminar flow in ducts. But the concept may be applied in a qualitative sense. For laminar duct flow, the Nu and f depend on the duct shape. Values of j_H/f and j_T/f for $Pr = 0.7$ (gases) are given in Table 6-4. As the Prandtl number increases, the j/f characteristic of laminar flow decreases, since $f \propto Pr^{-1/3}$ for laminar duct flows, and j/f is independent of Pr for turbulent flow.

Notes on Heat Transfer Augmentation

Augmentation is used to satisfy any of three objectives:

1. To obtain increased heat transfer duty for a given heat transfer surface area
2. To obtain reduced surface area and thereby achieve material savings and compactness
3. To achieve reduced pumping power for a given heat duty

The ratio of the thermal resistance between the two fluid streams is critical to assessing whether augmentation will be beneficial. Augmentation should be considered for the fluid stream that has the controlling thermal resistance. Conversely, if the tube side resistance is small, relative to the external resistance, the use of internally roughened tubes would have little effect on the overall heat transfer coefficient. If the thermal resistances of both streams are approximately equal, augmentation may be considered for both streams.

Augmentation can be limited by pressure drop, frontal area considerations, fouling potential, and the availability of the special surface geometries of the exchanger material.

The economic benefits of the augmented surface exchanger may be calculated after evaluating the performance improvement, accounting for the higher incremental cost of the special surface geometry. Augmentation is often not attractive unless it offers a cost advantage, relative to conventional technology.

Sample Calculation 6-1. A shell-and-tube exchanger is equipped with low integral-fins on the outer tube surface and has smooth inner surface tubes. The exchanger has 375 tubes, 0.650 inches inside diameter with 19 fins/inch by 10 feet in length. The tube side fluid is water ($Pr = 3$) and operates at $Re = 25,000$. Determine the potential for heat transfer augmentation.

Solution. The thermal resistance ratio between the shell-and-tube-side fluids must first be evaluated. Assume that this calculation yields $R_i R_o = \eta h_o A_o h_i A_i = 2$. Because the tube side resistance is 100% larger than on the shell side, tube-side augmentation should provide a substantial performance improvement. Possible choices for tube-side augmentation are internal fins, roughness, spiral tape inserts, or wire

Table 6-5
Equations for Film Condensation

$Re_L = 4\Gamma/\nu$		Properties evaluated at T_f
$F_1 = \dfrac{(k^3\rho^2 g)^{1/4}}{\mu}$	$F_2 = \dfrac{(k^3\rho^2 g)^{1/3}}{\mu}$	$T_f = T_{sat} + .75(T_w - T_{sat})$ $\Delta T = T_w - T_{sat}$

Vertical plate (T_w = const.)	
Laminar $Re_L < 1{,}800$	$h_{av} = .943\, F_1(h_{fg}/L\Delta T)^{1/4}$ or $h_{av}F_2 = 1.47\, Re_L^{-1/3}$
Turbulent	$h_{av} = .003 F_1^2(\Delta TL/\mu h_{fg})^{.5}$ $h_{av}F_2\mu^{1/3} = .0076\, Re_L^{0.4}$
Inclined plate	Replace g in vertical plate equations by g $\sin\theta$, where θ = angle from horizontal
On horizontal tube	$h = .725\, F_1(h_{fg}/D\Delta T)^{1/4}$
Row of N horizontal tubes	$h_N/h_1 = N^{-.25}$ h_N = average h over N Rows h_1 = average h on top tube
Inside vertical tube	$\dfrac{h}{C_{p\ell}G_m}\,Pr_\ell^{1/2} = .046\sqrt{\dfrac{\rho_{ff}}{\rho_v}}$ $G_m = \sqrt{(G_1^2 + G_1 G_2 + G_2^2)/3}$ G_1 = Mass velocity at top of tube G_2 = Mass velocity at bottom of tube
	f = Fanning friction factor evaluated at G_m
Inside horizontal tube	
Condensation of super-heated vapor	$q = \bar{h}A(T_{sat} - T_s)$ $\bar{h}$ = condensation coefficient for saturated vapor
Turbulent annular film	$h = 0.043\,\dfrac{k\rho^{1/2}}{\mu}Pr^{1/2}\tau_o^{1/2}$ τ_o = wall shear stress

coil inserts. Performance evaluation criteria involves calculating the percent increase of the heat transfer coefficient and the associated friction penalty. The preferred technique yields the highest heat transfer improvement per unit pressure drop, with due consideration given to cost, manufacturability, and fouling.

We may determine the performance improvement provided by internally roughened tubes. The design goal in this case, is to reduce the required heat transfer surface area. The most direct approach is tube replacement. The new exchanger will have the same 19 fins/inch extended surface and a two-dimensional helical rib roughness on the inner tube surface. For a direct tube replacement design, the internal roughness will provide a larger overall heat transfer coefficient, and less tube surface area will be required for equal heat duty (assuming constant water-side flow rate). The exchanger may have (1) a reduced number of tubes maintaining the 10 foot length, or (2) the same number of tubes but reduced length. Tube-side pressure-drop considerations suggest keeping the same number of tubes but reducing the length.

Assume that the internal roughness provides a 100% higher tube-side heat transfer coefficient and 150% greater friction factor than a smooth tube, both at Re = 25,000.

Using the Dittus-Boelter equation, the heat transfer coefficient of the smooth tube is:

$$h_{i,s} = \frac{.023k}{d} \, Re^{.8} Pr^{.4} = \frac{.023 \times .381}{0.650/12} \times 25,000^{.8} \times 3^{.4}$$

$$= 828 \ Btu/hr\text{-}ft^2\text{-}°F$$

The heat transfer coefficient on the external tube surface is:

$$h_{o,s} = h_{i,s} \frac{R_i}{R_o} \frac{A_i}{\eta A_o} = 828 \times 2 \times \frac{1}{.95 \times 3}$$

$$= 581 \ Btu/hr\text{-}ft^2\text{-}°F$$

The heat transfer coefficient of the rough tube is $2 \times 828 = 1,656$ Btu/hr-ft²-°F. The overall heat transfer coefficient (based on the tube-side surface area) of the smooth and augmented surface heat exchanger is:

$$\frac{1}{U_i} = \frac{1}{h_i} + \frac{1}{\eta_o h_o} \frac{A_i}{A_o}$$

For the smooth tube exchanger:

$$\frac{1}{U_{i,s}} = \frac{1}{828} + \frac{1}{.95 \times 602} \times \frac{1}{3} = 1.80 \times 10^{-3}$$

For the augmented design:

$$\frac{1}{U_{i,a}} = \frac{1}{1,656} + \frac{1}{.95 \times 602} \times \frac{1}{3} = 1.19 \times 10^{-3}$$

The augmented tube design provides a 51% higher overall heat transfer coefficient. Since $A' = U'A/U$, the augmented tube design requires only $2/3$ as much surface area. The tube side augmentation thus affords a 33% surface area reduction obtained by reducing the tube length from 10 feet to less than 7 feet.

The tube side pressure drop is computed next. Since the Reynolds number is the same for both designs:

$$\Delta P_a / \Delta P_s = (f/f_s)(L/L_s) = 2.5 \times 7/12 = 1.5$$

Although the circuit length is reduced by 33%, the higher friction factor of the augmented tubes causes a higher water-side pressure drop (about 50% more).

Surface Fouling

Heat exchanger surfaces become coated with deposits after some period of operation. This causes increased thermal resistance and a reduced U-value. Exchangers are therefore overdesigned to compensate for expected fouling. The fouling factor is defined as:

$$R_f = \frac{1}{U_{fouled}} - \frac{1}{U_{clean}} \qquad (6\text{-}22)$$

Fouling resistance causes greater performance reductions with liquids than with gases (mainly because heat transfer coefficients for liquids are higher).

Fouling resistance cannot be accurately predicted. Fouling factors are selected largely based on experience.

There are several types of fouling, namely:

- Crystallization fouling, which occurs when salts having an inverse solubility characteristic precipitate on a heat transfer surface hotter than the flowing fluid.
- Particulate fouling, which takes place when suspended solids deposit on the surface.
- Corrosion fouling, which occurs when the heat transfer surface material reacts with the fluid to yield corrosion deposits.
- Chemical reaction at the heat transfer surface (the surface material is not a reactant) may yield surface deposits. Examples are polymerization, cracking, and coking of hydrocarbons.
- Biofouling deposits, formed when biological mechanisms attach and grow on the heat transfer surface. Untreated water cooling systems are susceptible to biofouling.

- Freezing fouling, which occurs when the temperature of the heat transfer surface is below the solidification temperature of the fluid.

Fouling is a rate-dependent phenomena. Basic fouling theory assumes the net fouling rate is the difference between the solids deposition rate and their removal rate:

$$\frac{dM}{dt} = \dot{M}_d - \dot{M}_r \tag{6-23}$$

Depending on the magnitudes of the deposition and removal terms several fouling-rate characteristics are possible.

After an initial delay period, during which nucleation sites are established, deposits begin to accumulate. A linear growth of the fouling deposit occurs if the removal rate ($\dot{M}_r$) is small or if $\dot{M}_d$ and $\dot{M}_r$ are constant with $\dot{M}_d > \dot{M}_r$. The fouling resistance will attain an asymptotic value if $\dot{M}_d$ is constant and $\dot{M}_r$ is proportional to the deposition rate. The type of fouling characteristic depends on the fouling mechanism.

Fouling deposits may be removed by mechanical, chemical, or thermal cleaning. Outer surfaces of tubes may be cleaned with high pressure air, water, or steam. Inner tube surfaces are cleaned by forcing special plugs or brushes through the tubes. In most cases the exchanger must be taken out of service for cleaning. Chemical cleaning includes flushing the system with chemical cleaning agents selected to dissolve the particular fouling deposit. Thermal cleaning is accomplished by heating the exchanger to a sufficiently high temperature to vaporize the deposits. This method is limited in application and is mainly used for small gas-gas heat exchangers.

Closed water-cooling systems such as cooling towers and boilers employ chemical treatment to minimize the fouling characteristics of the water. See Reference 7 for details.

References

1. Shah, R. K., K. A. Afimiwala, and R. W. Mayne, "Heat Exchanger Optimization," *Heat Transfer 1978* (4):193–199 (1978).
2. Palen, J. W., T. P. Cham, and J. Taborek, "Optimization of Shell-and-Tube Exchangers by Case Study Method," AIChE, Reprint 1, 14th National Heat Transfer Conference, Atlanta, GA (1973).
3. Kays, W. M., and A. L. London, *Compact Heat Exchangers,* 2nd edition, McGraw-Hill Book Co., New York (1964).

4. Holman, J. P., *Heat Transfer,* 4th edition, McGraw-Hill Book Co., New York (1976).
5. Kern, D. Q., and A. D. Kraus, *Extended Surface Heat Transfer,* McGraw-Hill Book Co., New York (1972).
6. Colburn, A. P., *Ind. Eng. Chem.,* 35:873–877 (1933).
7. Cheremisinoff, N. P., and P. N. Cheremisinoff, *Cooling Towers: Selection, Design and Practice,* Ann Arbor Science Publishers, Ann Arbor, MI (1981).

7

HEADER DESIGN NOTES

Flow Distribution

In heat exchangers, the size of the frontal area on any side is based on the design and performance requirements. The size of piping for flows is based on the cost and pressure drop. Hence, the size of inlet and outlet piping is generally much smaller than the heat exchanger frontal area. Headers or manifolds are used to connect the supply and return pipes to the exchanger core.

An inlet header is a transition section that joins the exchanger's inlet face of the core or matrix to the supply line. The outlet header joins the outlet face of the core to the return pipe. Headers are referred to as a tank, box, or distributor (Figure 7-1).

A *manifold* is a flow channel in which the fluid enters at one end (or in the middle) and exits through side walls or vice versa.

Idealized heat transfer and pressure drop analyses assume that the fluid is uniformly distributed through the core. However, significant deterioration in performance can result when the flow distribution is nonuniform. A major consideration is to ensure that the fluid is distributed uniformly through the core. In design practices, care is taken to provide proper core face.

The inlet header design is most critical. Typical area increases from the inlet pipe to core face range from 5 to 50 times. It is not possible to maintain streamline flow with practical headers for such an area enlargement. A box configuration for the outlet header has an inlet header with a special contour to provide uniform flow distribution through the core, i.e., constant ΔP across the core. The flow continuously accelerates in the inlet header. For parallel flow headers, the inlet header dimension z_i can be either larger or smaller than the outlet dimension y_o; the theoretical shape of the inlet header is:

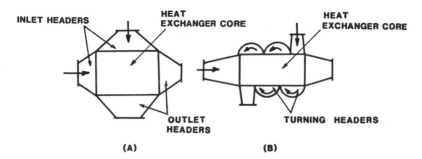

Figure 7-1. (A) Inlet and outlet, headers; (B) turning headers.

$$\frac{z}{y_o} = \frac{(1 - x^*)}{\left[\left(\frac{\rho_i}{\rho_o}\right)\left(\frac{\pi^2}{4}\right)x^{*2} + \left(\frac{y_o}{z_i}\right)^2\right]^{1/2}} \tag{7-1}$$

where ρ_i, ρ_o = fluid densities at the inlet of inlet header and outlet of exit header, respectively

x^* = X/L

The pressure drop in the inlet and outlet headers is:

$$\frac{\Delta P_i}{H_i} = 1 + 0.822 \left(\frac{H_o}{H_i}\right) \tag{7-2}$$

$$\frac{\Delta P_o}{H_i} = 0.645 \left(\frac{H_o}{H_i}\right) \tag{7-3}$$

$$\frac{\Delta P_t}{H_i} = 1 + 1.467 \left(\frac{H_o}{H_i}\right) \tag{7-4}$$

where $H_i = \left(\frac{\rho u_m^2}{2g_a}\right)_i$

$$H_o = \left(\frac{\rho u_m^2}{2g_c}\right)_o \tag{7-5}$$

In counterflow headers, the inlet and outlet headers are positioned on the same side of the exchanger core with the fluid flowing in the opposite direction.

A parallel flow configuration has a header loss (total-to-total), exclusive of core ΔP, equal to 2.47 times the inlet velocity head based on the velocity u_i. A free-discharge configuration has only one inlet velocity head loss and the counterflow configuration has 0.595 times the inlet velocity head loss or one-fourth that of the parallel-flow configuration. This means that a counterflow header arrangement is the preferred geometry. Of the total header loss, about 74% for the parallel arrangement and 50% for the counterflow arrangement is chargeable.

In normal-flow headers, the influent stream is perpendicular to the heat transfer core.

The inlet header acts as a diffuser due to a large flow-area increase from the inlet pipe to the exchanger core face.

There are four distinct flow regimes in a two-dimensional diffuser:

1. Unseparated flow or no appreciable stall
2. Transitory stall
3. Fully developed stall, where the flow separates from one wall and follows along the other
4. Jet flow where the flow separates from both walls

Figure 7-2 shows these flow regimes.

The ordinate scale in Figure 7-2 denotes the total included angle 2θ degrees of the diffuser, and the abscissa is the ratio of axial length N of a diffuser to the diffuser throat width W_1. Correlations for various lines a-a, b-b, etc., as well as the corresponding pressure recovery (rise) are available.[2]

If the inlet header is a box, a jet forms from the inlet pipe and will spread before impinging on the core face. Characteristics of spreading jets are described by Schlichting.[3] The included cone angle of a mixing turbulent jet is about 10°. To minimize the header volume and pressure losses, it is desirable to make the inlet header conical, and include a plenum chamber downstream (see Figure 7-3).

For liquid flows, a perforated baffle or perforated plate is used in the plenum chamber inlet header (see Figure 7-4). Header volume is considerably reduced and a fairly uniform flow distribution is attained at the core inlet with the use of a baffle or a perforated plate. For liquid flows, the core can be arranged such that the liquid enters at the bottom into the inlet header and flows vertically upward through the core. This in general assures a good flow distribution.

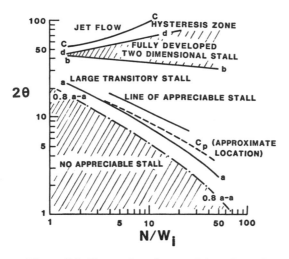

Figure 7-2. Flow regimes in a straight-walled diffuser.

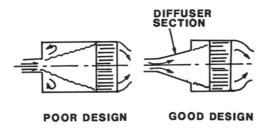

POOR DESIGN   **GOOD DESIGN**

Figure 7-3. Examples of designs to minimize header volume.

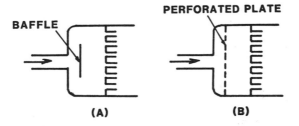

(A)   **(B)**

Figure 7-4. (A) Baffle; (B) perforated plate in the plenum chamber inlet header.

Considerations for Manifolds

There are four types of manifolds: dividing flow and combining-flow, and parallel and reverse-flow manifold systems. These are shown in Figure 7-5. Parallel- and reverse-flow systems are combinations of the basic dividing- and combining-flow manifolds interconnected by lateral branches or a heat exchanger core.

The pressure distribution in the dividing- and combining-flow headers and flow distribution through the lateral branches made up of circular tubes can be analyzed from the continuity and momentum equations for the header flows and the discharge equation for the lateral flows (see Bajura and Jones[4]). Some general conclusions about the pressure distribution in headers are:

1. More uniform flow distribution through the core is achieved through a reverse-flow manifold system than through the parallel-flow manifold system.
2. More uniform flow distribution through the core is achieved if the flow area of the combining flow header is larger than that of the dividing flow header.
3. Maintain the ratio of flow area (area of the pipe before the lateral branches) of the inlet header to flow area of lateral branches (exchanger core) greater than one to minimize flow maldistribution.
4. Try to maintain a significant pressure drop in the lateral branches (exchanger core) compared to the headers. This will provide uniform flow distribution through the core.

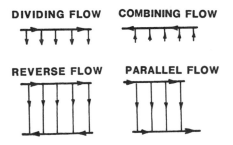

Figure 7-5. Types of manifolds.

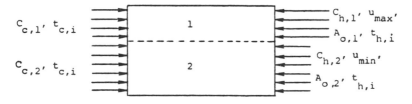

Figure 7-6. Idealized flow nonuniformity on one side of a counterflow heat exchanger.

Design Equations

Flow maldistribution within the core is due to nonuniformity at the core inlet face. This nonuniformity results from either upstream flow conditions or the shape of the inlet header. Gross blockage in the exchanger core can also cause maldistribution. Blockage may be due to brazing, soldering, or other manufacturing considerations.

It is possible to evaluate the reduction in the exchanger effectiveness due to nonuniform flow distribution on one side of the exchanger. Flow nonuniformity is made up of a two-step function velocity distribution (see Figure 7-6).

Figure 7-6 shows the case of a counterflow heat exchanger in which flow distribution is uniform on the cold side, and nonuniform on the hot side. The nonuniform flow distribution is represented by two different uniform velocities, u_{max} and u_{min}, flowing into the minimum free flow area $A_{o,1}$ and $A_{o,2}$, respectively.

The performance of the heat exchanger can be analyzed by considering two heat exchangers (1 and 2) in parallel, each of uniform velocity. The following analysis is based on the work of Carlson et al.[5], where relationships for the heat-capacity-rate ratios and number of transfer units for exchangers 1 and 2 of Figure 7-6 are given in terms of mean quantities for the whole exchanger as a single unit.

$$\frac{C_{h,1}}{C_h} = \frac{(\rho u_{max} A_{o,1} C_p)_h}{(\rho u_m A_o C_p)_h} = \left(\frac{u_{max}}{u_m}\right)\left(\frac{A_{o,1}}{A_o}\right)_h \tag{7-6}$$

$$\frac{C_{c,1}}{C_c} = \left(\frac{A_{o,1}}{A_o}\right)_c \tag{7-7}$$

$$\frac{C_{h,2}}{C_h} = \left(\frac{u_{min}}{u_m}\right)\left(\frac{A_{o,2}}{A_o}\right)_h \tag{7-8}$$

$$\frac{C_{c,2}}{C_c} = \left(\frac{A_{o,2}}{A_o}\right)_c \tag{7-9}$$

Note: $C_h = C_{h,1} + C_{h,2}$
$C_c = C_{c,1} + C_{c,2}$

We also define $D_h = 4A_oL/A$, where D_h and L are the same on each side of a counterflow exchanger.

$$\left(\frac{A_{o,1}}{A_o}\right)_h = \left(\frac{A_1}{A}\right)_h; \; \left(\frac{A_{o,2}}{A_o}\right)_h = \left(\frac{A_2}{A}\right)_h \tag{7-10}$$

$$\left(\frac{A_{o,1}}{A_o}\right)_h = \left(\frac{A_{o,1}}{A_o}\right)_c; \; \left(\frac{A_{o,2}}{A_o}\right)_h = \left(\frac{A_{o,2}}{A_o}\right)_c \tag{7-11}$$

where $A_o = A_{o,1} + A_{o,2}$
$A = A_1 + A_2$

The counterflow exchanger ϵ-N_{tu} relationship is valid for the C_{min} side (where N_{tu} = number of transfer units). This expression can be used for the C_h or C_c side as long as each quantity is defined consistently. The definition of each quantity on the C_h side is as follows. Note that C_h would be either C_{min} or C_{max}.

$$\epsilon_h = \frac{t_{h,i} - t_{h,o}}{t_{h,i} - t_{c,i}} \tag{7-12}$$

$$(N_{tu})_h = \frac{UA}{C_h} \tag{7-13}$$

where C^* is replaced by $\dfrac{C_h}{C_c}$ \hfill (7-14)

$$\epsilon_h = \frac{1 - \exp\left[-(N_{tu})_h(1 - C_h/C_c)\right]}{1 - (C_h/C_c)\exp\left[-(N_{tu})_h(1 - C_h/C_c)\right]} = \epsilon_{h,o} \tag{7-15}$$

Here the range of C_h/C_c is now $0 \le C_h/C_c \le \infty$. In this equation, $\epsilon_{h,o}$ is the temperature effectiveness on the hot side when the flow is uniform on *both* sides; C_h and C_c represent the heat capacity rates based on the total flow rate on the respective sides.

For nonuniform flow distribution on the hot side, the temperature effectiveness of exchangers 1 and 2 with proper subscripts is:

$$\epsilon_{h,1} = \frac{1 - \exp\left[-(N_{tu,1})_h(1 - C_{h,1}/C_{c,1})\right]}{1 - (C_{h,1}/C_{c,1}) \exp\left[-(N_{tu,1})_h(1 - C_{h,1}/C_{c,1})\right]} \tag{7-16}$$

A similar expression for $\epsilon_{h,2}$ can also be written. The following equations also apply:

$$\frac{C_{h,1}}{C_{c,1}} = \frac{C_{h,1}}{C_h}\frac{C_h}{C_c}\frac{C_c}{C_{c,1}} = \left(\frac{u_{max}}{u_m}\right)\left(\frac{C_h}{C_c}\right) \tag{7-17}$$

$$\frac{C_{h,2}}{C_{c,2}} = \left(\frac{u_{min}}{u_m}\right)\left(\frac{C_h}{C_c}\right) \tag{7-18}$$

$$(N_{tu,1})_h = \frac{UA_1}{C_{h,1}} \tag{7-19}$$

$$\frac{1}{UA_1} = \frac{1}{(\eta_o A_1 h_1)_h} + \frac{a'}{A_{w,1}k_w} + \frac{1}{(\eta_o h A_1)_c} \tag{7-20}$$

$(N_{tu,2})_h$ and $1/UA_2$ are defined by replacing the subscript 1 with 2.

On the hot side, the velocities in heat exchanger 1 and 2 are u_{max} and u_{min}, and would be different from the mean velocity u_m if the flow were completely uniform. This velocity change affects the heat transfer coefficient. Therefore, a proper heat transfer coefficient h_1 must be used for UA_1. Also, a proper h_2 must be used for the UA_2 expression. If fouling is considered, proper resistances must be added.

The exchanger's temperature effectiveness for the case of nonuniform flow distribution on the hot side is:

$$\epsilon_h = \frac{C_{h,1}\epsilon_{h,1} + C_{h,2}\epsilon_{h,2}}{C_h} \tag{7-21}$$

This definition is based on $q = q_1 + q_2$. The reduction in the temperature effectiveness on the hot side is:

$$\Delta\epsilon = \frac{\epsilon_{h,o} - \epsilon_h}{\epsilon_{h,o}} \tag{7-22}$$

$\epsilon_{h,o}$ is given by Equation 7-15. If the hot side is the C_{min} side, $\Delta\epsilon$ is the reduction in the exchanger effectiveness. If the cold side is the C_{min} side ($C^* = C_c/C_h$), $\Delta\epsilon$ of Equation 7-22 still represents the reduction in the exchanger effectiveness since:

$$\epsilon_h C_h = \epsilon_c C_c \qquad (7\text{-}23)$$

so that

$$\epsilon_c = \epsilon_h/C^*; \; \epsilon_{c,o} = \epsilon_{h,o}/C^* \qquad (7\text{-}24)$$

and

$$\Delta\epsilon = \frac{\epsilon_{c,o} - \epsilon_c}{\epsilon_c} = \frac{\epsilon_{h,o} - \epsilon_h}{\epsilon_{h,o}} \qquad (7\text{-}25)$$

This equation is based on flow nonuniformity on the hot side. For flow nonuniformity on the cold side, Equations 7-6 through 7-25 are applicable after interchanging subscripts h and c.

For the parallel-flow case with flow nonuniformity on one side, the procedure to determine the decrease in exchanger effectiveness is identical to that for the counterflow exchanger.

Another special case is a crossflow heat exchanger with nonuniform flow on the unmixed side, and uniform flow on the mixed side. The nonuniform flow distribution can be represented by two different uniform velocities on the hot side, and uniform flow distribution on the cold side (Figure 7-7). The maximum and minimum velocities are u_{max} and u_{min}, and the respective free flow areas are $A_{o,1}$ and $A_{o,2}$.

The performance of the exchanger can be analyzed by considering two heat exchangers (1 and 2) in parallel, each having uniform velocity.

The ϵ-N_{tu}-C^* relationship for the mixed or the unmixed side (as C_{min} can be represented by one equation in terms of C_h and C_c) is as follows:

$$\epsilon_h = \frac{C_c}{C_h}\left[1 - \exp\left\{-\frac{C_h}{C_c}\left[1 - \exp\left(-\frac{UA}{C_h}\right)\right]\right\}\right] \qquad (7\text{-}26)$$

The range of C_h/C_c is $0 \le C_h/C_c \le \infty$. In Equation 7-26, $\epsilon_{h,o}$ is the temperature effectiveness on the hot side when the flow is uniform on *both* sides; C_h and C_c are the heat capacity rates based on the total flow rate on respective sides. The temperature effectiveness ϵ_h is defined by Equation 7-12.

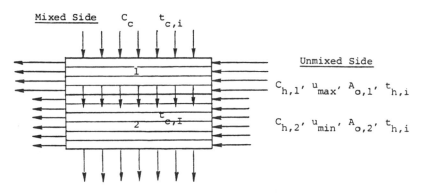

Figure 7-7. Idealized flow nonuniformity on the unmixed side of an unmixed-mixed crossflow heat exchanger.

For nonuniform flow distribution on the hot side (Figure 7-7), use the definition of Equation 7-12 for the temperature effectiveness of exchangers 1 and 2 with proper subscripts.

$$\epsilon_{h,1} = \frac{(t_{h,1} - t_{h,o})_1}{(t_{h,i} - t_{c,i})} \tag{7-27}$$

$$\epsilon_{h,2} = \frac{(t_{h,i} - t_{h,o})_2}{(t_{h,i} - t_{c,1})} \tag{7-28}$$

where $t_{c,1}$ is the outlet temperature on the cold fluid from exchanger 1.

$$\epsilon_{h,1} = \frac{C_c}{C_{h,1}} \left(1 - \exp\left\{ -\frac{C_{h,1}}{C_c}\left[1 - \exp\left(-\frac{UA_1}{C_{h,1}}\right)\right]\right\}\right) \tag{7-29}$$

$\epsilon_{h,2}$ can be obtained by replacing the subscript 1 with 2 in this equation. UA_1 is given by Equation 7-20 in which the proper heat transfer coefficient $(h_1)_h$ should be used. Also, a proper $(h_2)_h$ should be used in the UA_2 expression.

The temperature $t_{c,1}$ in Equation 7-28 is determined by the energy balance in exchanger 1. The heat transfer rate q_1 is:

$$q_1 = \epsilon_{h,1}C_{h,1}(t_{h,i} - t_{c,i}) = C_c(t_{c,1} - t_{c,i}) \tag{7-30}$$

so that

$$t_{c,1} = t_{c,i} + \frac{\epsilon_{h,1}C_{h,1}}{C_c}(t_{h,i} - t_{c,i}) \tag{7-31}$$

The heat transfer rate in exchanger 2 is:

$$q_2 = \epsilon_{h,2}C_{h,2}(t_{h,i} - t_{c,1}) \tag{7-32}$$

or

$$q_2 = \epsilon_{h,2}C_{h,2}\left(1 - \frac{\epsilon_{h,1} - C_{h,1}}{C_c}\right)(t_{h,i} - t_{c,1}) \tag{7-33}$$

The total heat transfer rate in the crossflow exchanger is:

$$q = q_1 + q_2 \tag{7-34}$$

The temperature effectiveness on the hot fluid side is:

$$q = \epsilon_h C_h(t_{h,i} - t_{c,i}) \tag{7-35}$$

The temperature effectiveness on the hot fluid side with the nonuniformity model is:

$$\epsilon_h = \frac{1}{C_h}\left[\epsilon_{h,1}C_{h,1} + \epsilon_{h,2}C_{h,2}\left(1 - \frac{\epsilon_{h,1}C_{h,1}}{C_c}\right)\right] \tag{7-36}$$

The reduction in the temperature effectiveness on the hot side is:

$$\Delta\epsilon = \frac{\epsilon_{h,o} - \epsilon_h}{\epsilon_{h,o}} \tag{7-37}$$

$\epsilon_{h,o}$ is given by Equation 7-26, where $\Delta\epsilon$ represents the reduction in the exchanger effectiveness.

This reduction in the exchanger effectiveness is based on flow nonuniformity on the hot side that was unmixed. For the case of flow nonuniformity on the cold side having unmixed flow and uniform flow on the hot side having mixed flow; the procedure is the same, except the subscripts h and c are interchanged. More complex situations are analyzed by Chiou.[6]

References

1. Ghose, S., and S. J. Kline, Report MD-36, Mechanical Eng. Dept., Stanford Univ., Stanford, CA (1976).
2. Kline, S. J., Mechanical Eng. Dept., "Report on Flow Phenomena and Design," Stanford Univ., Stanford, CA (July 1978).
3. Schlichting, H., *Boundary Layer Theory,* 6th edition, McGraw-Hill Book Co., New York (1968).
4. Bajura, R. A., and E. H. Jones, Jr., "Flow Distribution Manifolds," ASME Paper No. 76-FE-7 (1968).
5. Carlson, J. J., J. P. Johnston, and C. J. Sagi, *Trans. ASME, Journ. of Eng. for Power,* 90 (Series A):271–286 (1968).
6. Chiou, J. P., ASME Paper No. 77-WA/HT-3 (1977).

8

FLOW REGIMES AND
PRESSURE DROP IN
TWO-PHASE HEAT TRANSFER

Flow Regimes

Standard practice for heat exchanger analysis is to first identify the flow regimes and then employ the appropriate correlations.

Vertical Upward Cocurrent Flow

Flows of this type are shown in Figure 8-1

Bubbly flow. In this type, the gas or vapor phase is distributed as discrete bubbles in a continuous liquid phase. At one extreme, the bubbles may be small and spherical, and at the other extreme, the bubbles may be large with a spherical cap and a flat tail.

Slug flow. The gas or vapor bubbles are approximately the diameter of the pipe. The nose of the bubble has a characteristic spherical cap, and the gas in the bubble is separated from the pipe wall by a slowly descending liquid film. The liquid flow is contained in liquid slugs that separate successive gas bubbles. Slugs may or may not contain smaller entrained gas bubbles carried in the wake of the large bubble. The length of the main gas bubble varies.

Churn flow. Formed by the breakdown of the large vapor bubbles in slug flow. The gas or vapor flows chaotically through the liquid that is mainly displaced to the channel wall. The flow has a time-varying character and hence is called "churn flow." This region is also sometimes referred to as semi-annular or slug-annular flow.

Wispy-annular flow. The flow takes the form of a relatively thick liquid film on the walls of the pipe together with a considerable amount of liquid entrained in a central gas or vapor core. The liquid in the film is

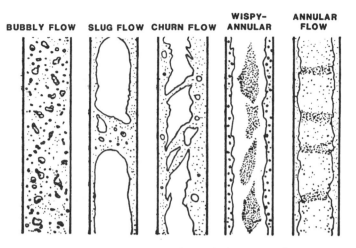

Figure 8-1. Flow patterns in vertical cocurrent flow.

aerated by small gas bubbles and the entrained liquid phase appears as large droplets which have agglomerated into long irregular filaments or wisps. This generally occurs at high mass velocities.

Annular flow. A liquid film forms at the pipe wall with a continuous central gas or vapor core. Large amplitude coherent waves are usually present on the surface of the film, and the continuous break up of these waves forms a source for droplet entrainment, which occurs in varying amounts in the central gas core.

Vertical Heated Channel Upward Flow

Heat flux through the channel wall alters the flow pattern from that which would have occurred in a long unheated channel at the same local flow conditions. These changes occur due to:

1. The departure from thermodynamic equilibrium coupled with the presence of radial temperature profiles in the channel.
2. The departure from local hydrodynamic equilibrium throughout the channel.

Figure 8-2 shows a vertical tubular channel heated by a uniform low heat flux and fed with liquid just below the saturation temperature.

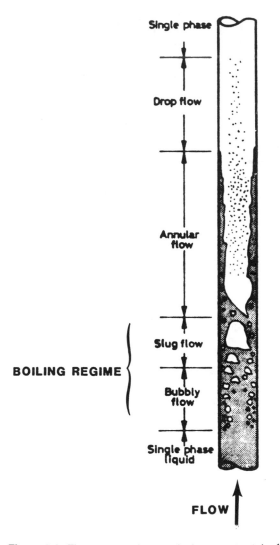

Figure 8-2. Flow patterns in a vertical evaporator tube.[1]

In the initial single-phase region, the liquid is heated to the saturation temperature. A thermal boundary layer forms at the wall, and a radial temperature profile forms. At some distance from the inlet, the wall temperature and the conditions for the formation of vapor (nucleation) at the wall are satisfied. Vapor forms at preferred positions on the tube surface. Vapor bubbles grow from these sites finally detaching to form a bubbly flow. With the production of more vapor, the bubble population increases with length and coalescence occurs, forming slug flow, which in turn gives way to annular flow further along the channel. Close to this point the formation of vapor at sites on the wall may cease and further vapor formation will result from evaporation at the liquid-film vapor-core interface. Increasing velocities in the vapor core cause entrainment of liquid in the form of droplets. The depletion of the liquid from the film by this entrainment and by evaporation finally causes the film to dry out completely. Droplets continue to exist and are slowly evaporated until only single-phase vapor is present.

Figure 8-3. shows the flow patterns of liquid-vapor flow in a heated pipe as a function of wall heat flux. Liquid enters the pipe at a constant flow rate and at a temperature lower than the saturation temperature. As the heat flux increases, the vapor appears closer and closer to the pipe inlet. The local boiling length is the extent of pipe where bubbles

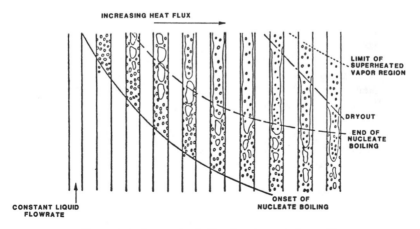

Figure 8-3. Convective boiling in a heated channel.[2]

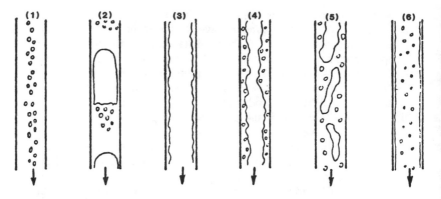

Figure 8-4. Air -water flow patterns in a downward cocurrent flow in a vertical pipe:[3] (1) bubbly, (2) slug, (3) falling film, (4) bubbly falling film, (5) churn, and (6) dispersed annular flow pattern.

form at the wall and condense in the liquid core where the liquid temperature is still lower than the saturation temperature. Vapor forms by:

1. Wall nucleation
2. Direct vaporization on the interfaces located in the flow itself

There is progressively less liquid between the wall and the interfaces. Consequently, the thermal resistance decreases along with the wall temperature, resulting in an end to wall nucleation. In annular flow, the liquid film flow rate decreases through evaporation and entrainment of droplets, although some droplets are redeposited. In heat flux controlled systems, when the film is completely dried out, the wall temperature rises very quickly and can exceed the melting temperature of the wall (called dryout). Flow patterns are shown in Figure 8-4.

In upward bubbly flow, bubbles are spread over the entire pipe cross section whereas in the downward flow bubbles gather near the pipe axis.

At higher gas flow rates (but a constant liquid flow rate) the bubbles agglomerate into large gas pockets. The tops of these gas plugs are dome-shaped whereas the lower extremity is flat with a bubbly zone underneath. This *slug flow* is generally more stable than in the upward case.

With annular flow, at small liquid and gas flow rates, a liquid film flows down the wall (*falling film flow*). If the liquid flow rate is higher, the bubbles are entrained within the film (*bubbly falling film*). At greater liquid and gas flow rates *churn flow* exists, which can evolve into dispersed annular flow for very high gas flow rates.

Horizontal Cocurrent Flow

The flow patterns for this type of flow are shown in Figure 8-5.

Bubbly flow (froth flow). This resembles the case in vertical flow except that the vapor bubbles tend to travel in the upper half of the pipe. At moderate gas and liquid velocities, the entire pipe cross-section contains bubbles. At higher velocities, a flow pattern equivalent to the wispy-annular pattern exists.

Plug flow. This is similar to slug flow in the vertical direction. Again, the gas bubbles tend to travel in the upper half of the pipe.

Stratified flow. This pattern only occurs at very low liquid and vapor velocities. The two phases flow separately with a relatively smooth interface.

Wavy flow. As the vapor velocity is increased, the interface becomes disturbed by waves traveling in the direction of flow.

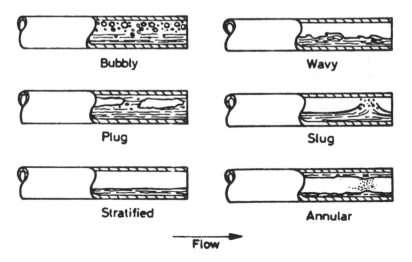

Figure 8-5. Flow patterns in horizontal flow.

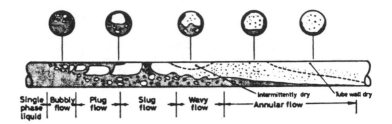

Figure 8-6. Flow patterns in a horizontal tube evaporator.[1]

Slug flow. At higher vapor velocities the waves at the interface break up to form a frothy slug which is propagated along the channel at a high velocity. The upper surface of the tube behind the wave is wetted by a residual film, which drains into the bulk of the liquid.

Annular flow. At higher vapor velocities a gas core forms with a liquid film around the periphery of the pipe. The film may or may not be continuous around the entire circumference but it will be thicker at the base of the pipe.

Flow patterns formed during the generation of vapor in horizontal tubular channels are influenced by departures from thermodynamic and hydrodynamic equilibrium. Figure 8-6 shows a horizontal tubular channel heated by a uniform low heat flux and fed with liquid just below the saturation temperature. The sequence of flow patterns corresponds to a relatively low inlet velocity (< 1 m/s). Note the intermittent drying and rewetting of the upper surfaces of the tube in wavy flow and progressive drying out over long tube lengths of the upper circumference of the tube wall in annular flow. At higher inlet liquid velocities, the influence of gravity is less obvious, the phase distribution becomes more symmetrical, and the flow patterns become closer to those in vertical flow.

Flow Normal to Tube Banks

The flow patterns in the crossflow zones are shown in Figure 8-7.

Spray flow. This occurs at high mass flow qualities with liquid carried along by the gas as a spray.

Bubbly flow. This occurs at low mass flow qualities with the gas distributed as discrete bubbles in the liquid.

Intermittent flow. Intermittent slugs of liquid are propelled cyclically by the gas.

Stratified-spray flow. The liquid and gas tend to separate with liquid flowing along the bottom. The gas-phase is entrained as bubbles in the liquid layer and liquid droplets are carried along by the gas as a spray.

Stratified flow. The liquid and gas are completely separated.

Spray and bubbly flows occur for either vertical up-and-down flow or horizontal side-to-side flow. Intermittent flow only occurs with vertical up-and-down flow and stratified-spray and stratified flow with horizontal side-to-side flow.

Flow Maps

A flow pattern map is a two-dimensional representation of the flow pattern existence domains. The respective patterns may be represented as areas on a graph, the coordinates of which are the actual superficial-phase velocities (j_ℓ or j_g). The coordinate systems are different according to various authors, and so far there is no agreement on the best coordinate system.

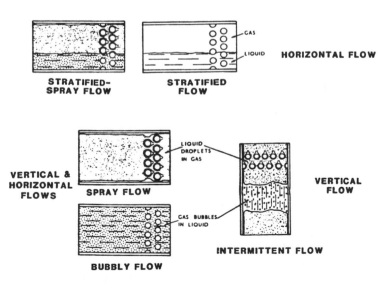

Figure 8-7. Shell-side two-phase flow patterns.[4]

Vertical Upward Flow

Figure 8-8 shows a flow pattern map based on observations on low-pressure air-water and high-pressure steam-water flow in small diameter (1–3 cm) vertical tubes.[5] The axes are the superficial momentum fluxes of the liquid ($\rho_\ell j_\ell$) and vapor ($\rho_g j_g^2$) phases, respectively. These superficial momentum fluxes can also be expressed in terms of mass velocity G and the vapor quality x:

$$\rho_\ell j_\ell^2 = \frac{[G(1-x)]^2}{\rho_\ell}; \quad \rho_g j_g^2 = \frac{(Gx)^2}{\rho_g} \tag{8-1}$$

Figure 8-8 should be considered as a rough guide only.

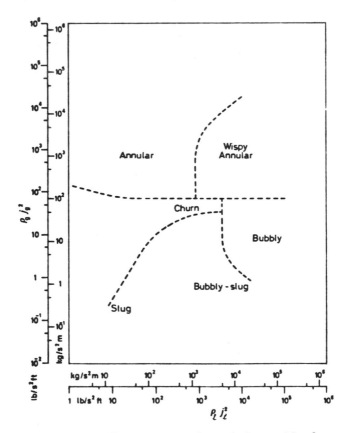

Figure 8-8. Flow pattern map for vertical upward flow.[5]

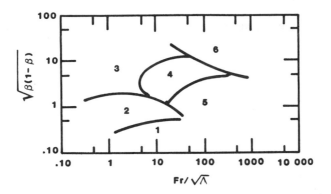

Figure 8-9. Flow pattern map for vertical downward flow:[3] (1) bubbly, (2) slug (3) falling film, (4) bubbly falling film, (5) churn, and (6) dispersed annular flow.

Vertical Downward Flow

Figure 8-9 shows one investigator's[3] chart. Data are based on two-component mixtures of air and different liquids flowing in a pipe 25.4 mm in diameter at a pressure of around 1.7 bar. The abscissa and ordinate are the quantities $Fr/\sqrt{\Lambda}$ and $\sqrt{\beta(1 - \beta)}$ (where β is the liquid holdup fraction) which are calculated at the test section pressure and temperature. The Froude number, Fr, is defined by:

$$Fr = (j_g + j_l)^2/gd_i \qquad (8\text{-}2)$$

where g = acceleration due to the gravity
 d_i = pipe diameter
 Λ = a coefficient that accounts for the liquid physical properties

$$\Lambda = (\mu/\mu_w)[(\rho_l/\rho_w)(\sigma/\sigma_w)^3]^{-1/4} \qquad (8\text{-}3)$$

where μ = liquid viscosity
 ρ = liquid density
 σ = liquid surface tension

Subscript w refers to water at 20°C and 1 bar.

Horizontal Flow

The well-known Baker plot[6] consists of a plot of G_g/λ and $G_\ell\lambda\psi/G_g$ where G_g and G_ℓ are the superficial mass velocities of the vapor and liquid phases, respectively. The factors λ and ψ are:

$$\lambda = \left[\left(\frac{\rho_g}{\rho_A}\right)\left(\frac{\rho_\ell}{\rho_w}\right)\right]^{1/2} \tag{8-4}$$

and

$$\psi = \left(\frac{\sigma_w}{\sigma}\right)\left[\left(\frac{\mu_\ell}{\mu_w}\right)\left(\frac{\rho_w}{\rho_\ell}\right)^2\right]^{1/3} \tag{8-5}$$

Baker's map has been modified by many investigators. Mandhane et al.[7] based a map upon 5,935 data points, 1,178 of which concern air-water flows. Its coordinates are the superficial velocities j_ℓ and j_g calculated at the test section pressure and temperature. The map is shown in Figure 8-10 and is valid for the parameter ranges given in Table 8-1.

Flow Normal to Tube Banks

Flow pattern maps for both vertical and horizontal flow normal to the tube banks are given in Figure 8-11. The parameters of these maps are those of Baker,[7] modified according to Bell et al.[8] It is a plot of superficial gas velocity vs. superficial liquid velocity with physical property terms attached. Superficial is used in the sense that the total flow area and not the actual phase flow area is used to evaluate the phase velocity. The flow area referred to is the minimum cross-sectional area for flow through the tube bank.

Estimating Pressure Drop

Two-phase drop in a shell-and-tube heat exchanger consists of friction, momentum change, and gravity.

$$\Delta P = \Delta P_f + \Delta P_m + \Delta P_g \tag{8-6}$$

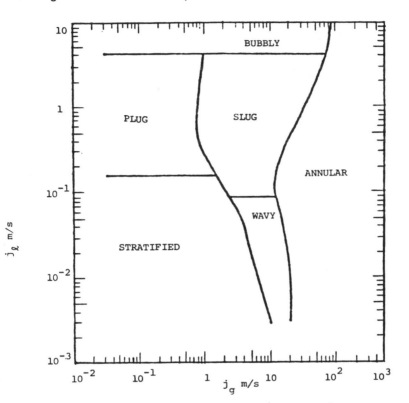

Figure 8-10. Flow map proposed by Mandhane et al.[7]

Table 8-1
Parameter Ranges for the Flow Map
Proposed by Mandhane et al.[7]

Conditions	Range of Values	
Pipe inner diameter	12.7 – 165.1	mm
Liquid density	705 – 1,009	$kg\text{-}m^{-3}$
Gas density	0.80 – 50.5	$kg\text{-}m^{-3}$
Liquid viscosity	$3 \times 10^{-4} - 9 \times 10^{-2}$	Pa
Gas viscosity	$10^{-5} - 2.2 \times 10^{-5}$	Pa
Surface tension	24 – 103	$mN\text{-}m^{-1}$
Liquid superficial velocity	0.09 – 731	$cm\text{-}s^{-1}$
Gas superficial velocity	0.04 – 171	$m\text{-}s^{-1}$

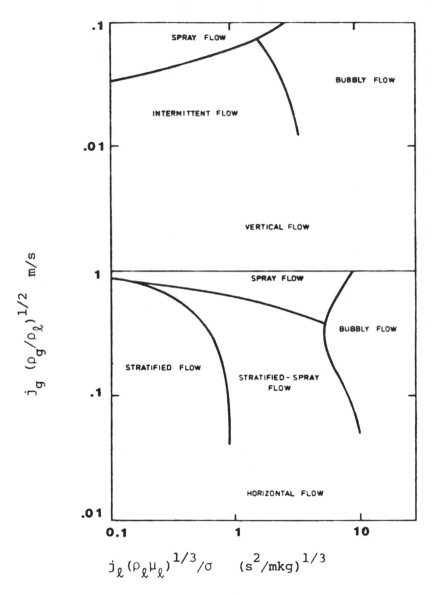

Figure 8-11. Shell-side flow pattern maps.[4]

The entrance and exit pressure losses, usually considered in a compact heat exchanger application, are neglected because of

1. The lack of two-phase data for these pressure losses
2. Their small contribution to the total pressure drop for tubular exchangers

The evaluation of ΔP due to momentum and gravity effects is generally based on a homogeneous model.

Homogeneous Flow Model

This is the simplest two-phase flow model. The basic premise is that a real two-phase flow can be replaced by a single-phase flow with the density of the homogeneous mixture defined by:

$$\nu_{hm} = \nu_\ell(1 - x) + \nu_g x \qquad (8\text{-}7A)$$

$$\frac{1}{\rho_{hom}} = \frac{1 - x}{\rho_\ell} + \frac{x}{\rho_g} \qquad (8\text{-}7B)$$

where ν is the specific volume. Subscripts ℓ and g denote liquid and gas phases and x is the quality (the ratio of gas mass flow rate to total (gas + liquid) mass flow rate).

The pressure drop/rise due to an elevation change is:

$$\Delta P_g = \pm \rho_{hom}\left(\frac{g}{g_c}\right)L \sin \theta \qquad (8\text{-}8)$$

Angle θ is measured from the horizontal. The $+$ sign stands for a downflow, and the $-$ sign stands for an upflow. Gravity pressure drop predictions from this theory are good for high quality and high pressure applications. When ΔP_g is predominant (one half to two thirds of the ΔP), such as for low velocities and low pressure applications, the following equation, which takes into account the velocity slip between two phases via the void fraction α (the ratio of gas volume to total volume), should be used:

$$\Delta P_g = \pm (\rho_\ell(1 - \alpha) + \rho_g\alpha)(g/g_c)\, L \sin \theta;$$

for $\Delta P_g > 0.5\Delta P_{total}$ $\qquad (8\text{-}9)$

The momentum pressure drop/rise from the homogeneous model is:

$$\Delta P_m = \frac{G^2}{g_c}\left(\frac{1}{\rho_2} - \frac{1}{\rho_1}\right) \tag{8-10}$$

ρ_2 and ρ_1 are the densities of homogeneous mixtures at the exchanger (tube) outlet and inlet, respectively. They are individually evaluated using Equation 8-7. G is the mass velocity.

Separated Flow Model

Here, the two phases are artificially segregated into two streams. Each stream (vapor and liquid) is under the same pressure gradient but not necessarily with the same velocity. The separated flow model reduces to the homogeneous flow model if the mean velocities of the two streams are the same. The best known separated flow model is the Lockhart and Martinelli[9] correlation. In the Lockhart-Martinelli method, the two fluid streams are considered segregated. The conventional pressure-drop friction-factor relationship is applicable to individual streams. The liquid- and gas-phase pressure drops are considered equal irrespective of the flow patterns. ϕ_ℓ^2 denotes the ratio of a two-phase frictional pressure drop to a single-phase frictional pressure drop for the *liquid* flowing alone in the tube.

$$\phi_\ell^2 = \frac{\Delta P_f}{\Delta P_\ell} \tag{8-11}$$

And for vapor:

$$\phi_g^2 = \frac{\Delta P_f}{\Delta P_g} \tag{8-12}$$

where ΔP_g is the single-phase frictional pressure drop for the *gas* flowing alone in the tube.

χ^2 is the ratio of a single-phase pressure drop for the liquid phase flowing alone in the tube to that for the gas phase flowing alone in the tube.

$$\chi^2 = \frac{\Delta P_\ell}{\Delta P_g} \tag{8-13}$$

The correlation is shown in Figure 8-12 and the curves can be represented in equation form as:

$$\phi_\ell^2 = 1 + \frac{C}{\chi} + \frac{1}{\chi^2}$$

(8-14)

or

$$\phi_g^2 = 1 + C\chi + \chi^2$$

(8-15)

where the value of C is dependent upon the four possible single-phase flow regimes.

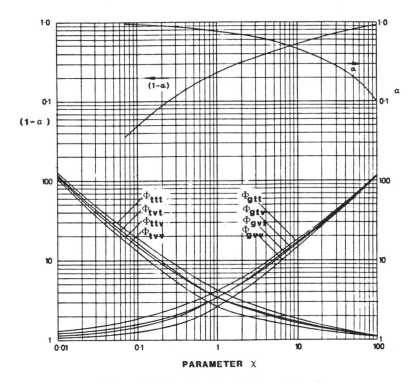

Figure 8-12. Lockhart-Martinelli correlation.[9]

Liquid		Gas		C
Turbulent	-	Turbulent	(tt)	20
Viscous	-	Turbulent	(vt)	12
Turbulent	-	Viscous	(tv)	10
Viscous	-	Viscous	(vv)	5

$$(8\text{-}16)$$

The two-phase frictional pressure drop by the Lockhart-Martinelli method is determined as follows. First, from the amount of liquid and gas flow rates, and using corresponding friction factors or appropriate correlations, ΔP_ℓ and ΔP_g are calculated. The liquid flow is considered to occupy the entire cross-section for the ΔP_ℓ evaluation, and the gas flow occupies the whole cross-section for the ΔP_g evaluation. The parameter χ is then calculated from Equation 8-13. The value of C is determined from Equation 8-16 and ϕ_ℓ or ϕ_g are computed from Equations 8-14 and 8-15. The two-phase frictional pressure drop is then calculated from the definition of ϕ.

The Lockhart-Martinelli method was developed for two component adiabatic flows at a pressure close to atmospheric. Martinelli and Nelson[10] extended this method for forced convection boiling for all pressures up to the critical point. The mixture of steam and water was considered "turbulent-turbulent." They presented ϕ_{fo}^2 graphically as a function of the quality x and the system pressure as shown in Figure 8-13.

$$\phi_\ell^2 = \frac{\Delta P_f}{\Delta P_{\ell o}}$$

$$(8\text{-}17)$$

where $\Delta P_{\ell o}$ is the frictional pressure drop for the liquid flow alone, in the same tube, with a mass flow rate equal to the *total* mass flow rate of the two-phase flow.

The Martinelli-Nelson experimental curves of ϕ_{fo} vs. x show breaks in the slope due to changes in flow regimes. Surface tension is not included although it may have a significant influence at high pressure near the critical point. The Martinelli-Nelson method provides more correct results than the homogeneous model for low mass velocities (G < 1,360 kg/m²s). In contrast, the homogeneous model provides better results for high mass velocities.

Chisholm[11] gives the following correlation for flow of evaporating two-phase mixtures that accounts for some of the effects neglected in other methods.

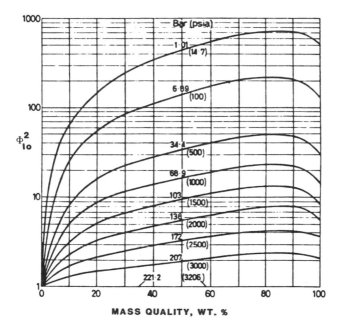

Figure 8-13. Martinelli-Nelson correlation.[10]

$$\phi_{\ell o}^2 = \frac{\Delta P_f}{\Delta P_{\ell o}}$$

$$= 1 + (\Gamma^2 - 1)[Bx^{(2-n)/2}(1 - x)^{(2-n)/2} + x^{2-n}] \qquad (8\text{-}18)$$

where
$$\begin{aligned}
B &= (C\Gamma - 2^{2\text{-}n} + 2)/(\Gamma^2 - 1) & (8\text{-}19) \\
\Gamma^2 &= \Delta P_{go}/\Delta P_{\ell o} & (8\text{-}20) \\
C &= (\rho_\ell/\rho_g)^{1/2}/K + K(\rho_g/\rho_\ell)^{1/2} & (8\text{-}21) \\
K &= \text{velocity ratio} = j_g/j_\ell & (8\text{-}22)
\end{aligned}$$

n is the exponent in the Blasius relation for friction factor $f = C_1/Re^n$, with $n = 0.25$ for the turbulent flow. These discussions are inclusive of tube flow only.

Two-phase pressure-drop correlations for the shell-side flow are available for a segmentally baffled shell-and-tube exchanger. The fric-

Table 8-2
Values of B for Two-Phase Frictional Pressure-Drop
Evaluation in Crossflow and Window-Flow
Zones by Equation 8-18

Zone	Horizontal	Vertical Up and Down Flow
Crossflow		
Spray and bubble	0.75	1.0
Stratified and Stratified spray	0.25	—
Window (n = 0)	$2/(\Gamma + 1)$	$(\rho_\ell/\rho_{hom})^{0.25}$

tional pressure drop consists of two-components, one associated with the crossflow zone and the other with the window zone. Grant and Chisholm[4] determined the components of the pressure drop. The two-phase crossflow zone and window zone frictional pressure drops are given by Equation 8-18 with values of B given in Table 8-2. Values of exponent n for the crossflow zone are: n = 0.46 for horizontal side-to-side flow, and n = 0.37 for vertical up-and-down flow.

References

1. Collier, J. G., *Convective Boiling and Condensation*, McGraw-Hill Book Co., New York (1972).
2. Hewitt, G. F., and N. S. Hall-Taylor, *Annular Two-Phase Flow*, Pergamon Press, London (1970).
3. Oshinowo, T., and M. E. Charles, *Can. Journ. of Chem. Engrg.*, 52: 25–35 (1974).
4. Grant, I. D. R., and D. Chisholm, *Trans. ASME, Journal of Heat Transfer*, 101 (Series C): 38–42 (1979).
5. Hewitt, G. F., and D. N. Roberts, "Studies of Two-Phase Flow Patterns by Simultaneous X-Ray and Flash Photography," AERE-M2159, H.M.S.O. (1969).
6. Baker, O., *Oil and Gas Journ.*, 53(12): 185–190 (1954).
7. Mandhane, J. M., G. A. Gregory, and K. Aziz, *Intl. Journ. of Multi Flow*, 1: 537–553 (1974).

8. Bell, K. J., J. Taborek, and F. Fenoglio, *Chem. Engrg. Progress Symposium Series*, (Heat Transfer—Minneapolis), 66 (102): 150–165 (1970).
9. Lockhart, R. W., and R. C. Martinelli, *Chem. Engrg. Prog.*, 45: 39–48 (1949).
10. Martinelli, R. C., and D. B. Nelson, *Trans. ASME*, 70: 695 (1948).
11. Chisholm, D., *Intl. Journ. of Heat and Mass Transfer*, 16: 347–358 (1973).

9

CORRELATIONS FOR CONDENSATION

General Notes

Condensation is a process whereby heat is removed from a system in such a manner that vapor is converted into liquid. This occurs when vapor is cooled sufficiently below the saturation temperature to induce the nucleation of droplets. Nucleation may occur *homogeneously* within the vapor or *heterogeneously* on entrained particulate matter. Heterogeneous nucleation may also occur on the walls of the system, particularly if these are cooled (an example is that of a surface condenser). The forms of heterogeneous condensation are dropwise and filmwise. Filmwise condensation occurs on a cooled surface that is easily wetted. On nonwetted surfaces, the vapor condenses as drops that grow by further condensation and coalescence and then roll over the surface. New droplets then form to take their place.

Notes on Dropwise Condensation

This condensation occurs on nonwettable surfaces. The condensate appears in the form of droplets that grow and coalesce with adjacent droplets. Subsequently, they fall from the surface due to gravity or aerodynamic drag forces. New droplets then appear on the exposed surfaces.

There are two basic models that describe dropwise condensation. The first postulates that condensation initially occurs in a filmwise manner on a thin unstable liquid film covering the surface. On reaching a critical thickness, the film ruptures and the liquid is drawn into droplets by surface tension forces. This process continually repeats itself. A quantitative treatment of dropwise condensation based on radially inward drainage of condensate towards a droplet is given by Silver,[1] where the ratio between dropwise and filmwise condensation rates are given as:

$$\frac{j_D}{j_F} = \left[\frac{\rho_\ell^2 D^2 g}{24.2\mu_f j_F}\right]^{1/9} \tag{9-1}$$

where j_D = dropwise condensation mass flux

j_F = filmwise condensation mass flux under identical conditions

For steam at atmospheric pressure having $j_F = 1.36 \times 10^{-2}$ kg/m², this ratio is 6.5, and the mean thickness of the thin film is about 2 μm.

The second model assumes that the droplet formation is a heterogeneous nucleation process. In general, the state-of-the-design art for dropwise condensation is poor, with considerable contradictions in the literature. The droplet nucleation model appears more likely at low condensation rates (i.e., temperature differences up to 5°C). At higher condensation rates, there may be a film disruption mechanism as an intermediate stage before establishing fully developed filmwise condensation. The transition to filmwise condensation at high temperature differences is shown in Figure 9-1.

Notes on Filmwise Condensation

In this type of condensation, the condensate wets the surface and forms a continuous liquid film. The hydrodynamics of film flow are well known; gravity, vapor shear, and surface tension forces, either singly or together, serve to promote removal of the condensate film. Two major condensation mechanisms are the gravity-controlled (vapor space) condensation and vapor-shear-controlled (forced convective) condensation.

Figure 9-2 summarizes the principal resistances to heat transfer during condensation. For a pure, saturated vapor, the resistance at the vapor/liquid interface is small and, to a first approximation, may be neglected.

For a superheated vapor containing a noncondensible gas or a multicomponent vapor mixture, significant resistance to both heat and mass transfer at the vapor/liquid interface exists. Mass and heat transfer processes occur in parallel within the gas phase and must be considered together to establish the respective driving forces.

Heat transfer also takes place through the condensate liquid film to the wall and then to the coolant. With a pure saturated vapor, the tem-

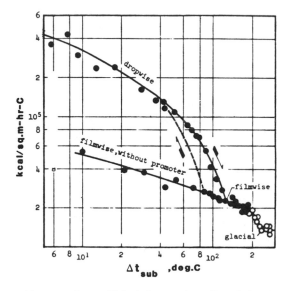

Figure 9-1. Heat transfer coefficients for condensation of steam on a short vertical copper surface (palmitic acid used as promoter for dropwise condensation).[2]

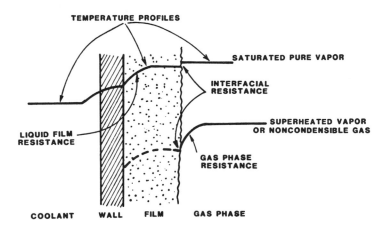

Figure 9-2. Principal resistances to heat transfer during condensation.

perature drop across the liquid film represents the major resistance to heat transfer. Techniques that reduce the condensate film thickness or promote a higher "effective" conductivity will increase the condensing-side heat transfer coefficient. With a vapor containing noncondensibles or with a multicomponent vapor mixture, the gas-phase resistance often dominates the overall condensing-side coefficient, and increases in the condensate film coefficient may not promote a comparable increase in the overall condensation rate.

Correlations for Filmwise Condensation

The following paragraphs summarize film condensation correlations for frequently encountered geometries.

Film Condensation on a Planar Surface

Nusselt[3] analyzed this configuration based on the following assumptions:
1. The condensate film is laminar
2. The fluid properties are constant
3. Subcooling of the condensate film is negligible
4. The momentum changes through the condensate film are negligible
5. The vapor is stationary and exerts no drag on the downward motion of the condensate (i.e., the condensation is gravity controlled)
6. Heat transfer is by conduction only

For a surface inclined at an angle θ to the horizontal, the following equation can be used to estimate the mean heat transfer coefficient.

$$h_m\left[\frac{\mu_\ell^2}{k_\ell^3\rho_\ell(\rho_\ell - \rho_g)g}\right]^{1/3} = 1.47\ Re_\Gamma^{-1/3} \tag{9-2}$$

$$\text{where}\ \ Re_\Gamma = \frac{4\Gamma_z}{\mu_\ell} \tag{9-3}$$

$$\Gamma_z = \frac{\rho_\ell(\rho_\ell - \rho_g)g\sin\theta\ \delta^3}{3\mu_\ell} \tag{9-4}$$

$$\delta = \left[\frac{4\mu_\ell k_\ell z(t_{g,i} - t_w)}{\rho_\ell(\rho_\ell - \rho_g)g \sin\theta \, i_{\ell g}} \right]^{1/4} \qquad (9\text{-}5)$$

where Γ_z = condensate mass flow rate per unit width of the planar surface at a distance z from the top of the plane surface

δ = condensate film thickness at the distance z

$i_{\ell g}$ = latent heat of vaporization

$t_{g,i}$ = the film surface (interface) temperature

t_w = wall temperature

The local heat transfer coefficient for the laminar condensate film is:

$$h_z = \frac{k_\ell}{\delta} \qquad (9\text{-}6)$$

The local heat transfer coefficient for a turbulent condensate film over a vertical surface is given by the Colburn[4] equation:

$$\frac{h_z}{k_\ell} \left[\frac{\mu^2_\ell}{\rho_\ell(\rho_\ell - \rho_g)g} \right]^{1/3} = 0.056 \, \mathrm{Re}_\Gamma^{0.2} \mathrm{Pr}_\ell^{1/3} \qquad (9\text{-}7)$$

Condensation over Horizontal Tubes

The following correlations apply to gravity-controlled condensation. The mean heat transfer coefficient for laminar film condensation over a single horizontal tube is:[3]

$$h_m \left[\frac{\mu^2_\ell}{k^3_\ell \rho_\ell(\rho_\ell - \rho_g)g} \right]^{1/3} = 1.51 \, \mathrm{Re}_\Gamma^{-1/3} \qquad (9\text{-}8)$$

The same h_m will exist for vertical and horizontal tubes in stagnant vapor for the same temperature difference and same fluid conditions at

$$z = 2.87 \, d_o \qquad (9\text{-}9)$$

In a bank of horizontal tubes, the condensate runs off the bottom of the upper tube onto successive tubes below. The mean heat transfer coefficient for n tubes to that of the first (top) tube is (from Kern):[5]

$$\frac{h_m}{h_{m,1}} = n^{-1/6} \qquad (9\text{-}10)$$

The ratio of the mean heat transfer coefficient for the nth tube to the first tube is:

$$\frac{h_{m,n}}{h_{m,1}} = n^{5/6} - (n - 1)^{5/6} \tag{9-11}$$

An analytical expression accounting for the influence of vapor shear on the heat transfer coefficient for flow across a horizontal tube follows.[6] The model assumes no separation of the laminar vapor boundary layer and laminar flow within the condensate film.

$$\frac{h_m d_o}{k_\ell} = 0.9 \left[\frac{\rho_\ell u_{g\infty} d_o}{\mu_\ell}\right]^{0.5} \tag{9-12}$$

The crossflow is downward over the tube per Equation 9-12. $u_{g\infty}$ = vapor velocity approaching the tube. At high values of the Reynolds number ($> 10^6$), separation of the boundary layer occurs and Equation 9-12 overpredicts. At these high Reynolds values, the proper equation is the same as Equation 9-12 but with the 0.9 replaced by a factor 0.59. Note that condensation tends to inhibit boundary layer separation. For lower velocities where both gravitational and shear forces are important the following equation holds:

$$h_m = \left\{\frac{1}{2}h_{shear}^2 + \left[\frac{1}{4}h_{shear}^4 + h_{grav}^4\right]^{1/2}\right\}^{1/2} \tag{9-13}$$

h_{shear} and h_{grav} are defined as before.

Condensation Inside a Vertical Tube

Laminar film condensation inside a vertical tube with negligible vapor shear can be described by the planar surface case (see Equation 9-2). At high velocities, where the flow is vapor-shear controlled, the following correlation is recommended[7] for the local heat transfer coefficient:

$$\frac{h_z \mu_\ell}{k_\ell \rho_\ell^{1/2}} = 0.036 \, Pr_\ell^{0.65} \tau_w^{1/2} \tag{9-14}$$

where τ_w is the shear stress at the outer edge of the laminar sublayer.

Condensation Inside a Horizontal Tube

The system of condensation within a horizontal tube is covered in detail by Bell et al.[8] and Collier.[9] Correlations for the annular flow regime in a horizontal tube are the same as those for a vertical tube except that the gravity contribution in τ_w is zero. These correlations are given by Collier.[9]

For the stratified-flow case the correlation that follows is recommended. The stratified flow regime corresponds to a regime of small vapor velocity and low interfacial shear forces. The Nusselt equation (Equation 9-8) for condensation on the outside of a horizontal channel is applicable in a modified form. The system geometry for this situation is shown in Figure 9-3. As shown, the laminar film of condensate runs down the inside upper surface of the tube and collects as a stratified fluid layer in the lower portion of the tube. Equation 9-8 is modified by accounting for the effect of tube inclination on the rate of condensation.

$$h_m = \frac{1}{\pi}\left[\frac{\rho_\ell(\rho_\ell - \rho_g)k_\ell^3 i_{\ell g}g}{1.5d_i\mu_\ell\Delta t}\right]^{1/4} F(\phi)(\cos\theta)^{1/4} \qquad (9\text{-}15)$$

Note: h_m is based on the entire inside wall surface area. θ is the tube inclination from the horizontal. Parameter $F(\phi)$ is defined as follows:

$$F(\Omega) = \left[\frac{4}{3}\int_0^\Omega \sin^{1/3}\phi\,d\phi\right]^{3/4} \simeq 0.81\Omega \qquad (9\text{-}16)$$

Note that there is a pressure gradient and the condensate at the tube outlet often fills the tube cross-section. That is, the hold-up angle 2θ

Figure 9-3. Laminar condensation inside a horizontal tube.

($= 2\pi - 2\Omega$) increases along with the tube length. Rufer et al.[10] gives an expression for its value.

References

1. Silver, R. S., *Proceedings of the Institution of Mechanical Engineers*, 178 (Part 1, No. 14): 339–376 (1964).
2. Takeyama, T., and S. Shimizu, *Heat Transfer 1974*, 111:274–278 (1974).
3. Nusselt, W., *Die Oberflachenkondensation des Wasserdampfer*, *VOL Z*, 60:541, 569 (1916).
4. Colburn, A. P., *Trans. AIChE*, 30:187–193, (1933–34).
5. Kern, D. Q., *Process Heat Transfer*, McGraw-Hill Book Co., New York (1950).
6. Shekriladze, I. G., and V. I. Gomelauri, *Intl. Journ. of Heat and Mass Transfer*, 9:581–591 (1966).
7. Soliman, M., J. R. Schuster, and P. J. Berenson, *Journ. of Heat Transfer*, 90:267–276 (1968).
8. Bell, K. J., J. Taborek, and F. Fenoglio, *Chem. Engrg. Progress Symposium Series*, (Heat Transfer-Minneapolis), 66 (102): 150–165 (1970).
9. Collier, J. G., *Convective Boiling and Condensation*, McGraw-Hill Book Co., New York (1972).
10. Rufer, C. E., and S. P. Kezios, *Journ. Heat Transfer*, 88:265–275 (1966).

10

PRINCIPLES OF BOILING

General Notes on Pool Boiling

Pool boiling takes place on a heated surface submerged in a large volume of stagnant liquid. When the liquid is at its boiling point, it is referred to as *saturated pool boiling*. When it is below the boiling point, the phenomenon is called *subcooled pool boiling*. Heat transfer rates in pool boiling are plotted on logarithmic coordinates as heat flux per unit area versus wall temperature (or wall superheat). Figure 10-1 gives an example of saturated pool boiling of degassed liquid. Such a boiling curve can be traced out entirely by heaters maintaining a constant wall temperature (high temperature fluid or condensing vapor as the heat source) or partially with constant-heat flux heaters (electric heating). The various stages of pool boiling shown by the boiling curve are illustrated in Figure 10-2.

When a constant-heat flux system is used to generate a boiling curve, a first-order instability develops as the power is increased and the operating point shifts rapidly to the film boiling region. This new operating point corresponds to a temperature greater than the melting temperatures; hence, point D is commonly referred to as the burnout point. If operation in the film boiling region is obtained, the power may be increased to actual burnout or reduced to the Leidenfrost point, where the system reverts back to operation in the nucleate region.

Saturated Pool Boiling Correlations

Summaries of heat transfer correlations for the various regions of the pool boiling curve in Figure 10-1 follow.

For *single-phase natural convection* the correlations are of the form:

$$Nu_f = f(Gr, Pr) \tag{10-1}$$

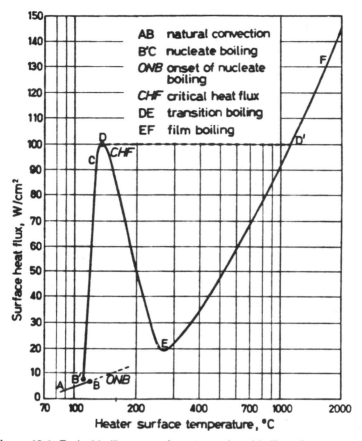

Figure 10-1. Typical boiling curve for saturated pool boiling of water at atmospheric pressure.[1]

where Nu_f = mean Nusselt number
    Gr = Grashof number
    Pr = fluid Prandtl number

All the fluid properties in Nu_f, Gr, and Pr are normally evaluated at the film temperature $(T_w + T_m)/2$.

For the geometries of a vertical flat plate and a vertical cylinder, the mean Nusselt number is a function of Rayleigh number Ra = GrPr. Table 10-1 gives typical values.

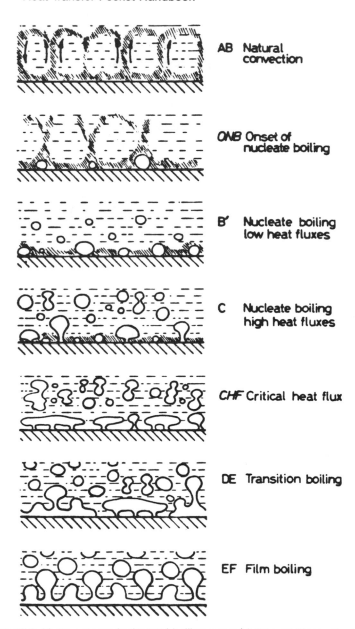

AB Natural convection

ONB Onset of nucleate boiling

B′ Nucleate boiling low heat fluxes

C   Nucleate boiling high heat fluxes

CHF Critical heat flux

DE Transition boiling

EF Film boiling

Figure 10-2. Various stages in the pool boiling curve.[1] Letter notations refer to various positions of the boiling curve in Figure 10-1.

Results are approximated by an equation in the indicated range of Rayleigh number:

$$Nu_f = 0.59\ Ra^{1/4};\ 10^4 \le Ra \le 10^9$$

$$Nu_f = 0.13\ Ra^{1/3};\ 10^9 \le Ra \le 10^{12} \tag{10-2}$$

The characteristic size of Nu_f is length L for a vertical plate, and the outside diameter d_o for a circular cylinder.

For horizontal cylinders, the mean Nusselt number is listed in Table 10-2 as a function of Ra.

For $10^3 \le Ra \le 10^9$, the mean Nusselt number is:

$$Nu_f = 0.53\ Ra^{0.25} \tag{10-3}$$

where the characteristic dimension is the cylinder's outside diameter.

The following equations apply to horizontal heated plates facing upward or cooled plates facing downward:[1]

$$Nu_f = 0.54\ Ra^{1/4};\ 10^5 \le Ra \le 2 \times 10^7$$

Table 10-1
Mean Nusselt Number for Natural Convection Over Vertical Flat Plates and Vertical Cylinders[1]

Ra	Nu_f	Ra	Nu_f	Ra	Nu_f
1	1.44	10^4	6.03	10^8	58.9
10	1.90	10^5	10.5	10^9	105.
10^2	2.63	10^6	18.6	10^{10}	214.
10^3	3.89	10^7	33.1	10^{11}	501.

Table 10-2
Mean Nusselt Number for Natural Convection Over Horizontal Cylinders[1]

Ra	Nu_f	Ra	Nu_f	Ra	Nu_f
1	1.08	10^3	3.16	10^6	16.2
10	1.51	10^4	5.37	10^7	28.8
10^2	2.11	10^5	9.33	10^8	51.3
				10^9	93.3

$$Nu_f = 0.14 \, Ra^{1/3}; \quad 2 \times 10^7 \leq Ra \leq 3 \times 10^{10} \tag{10-4}$$

For a horizontal heated plate facing downward or cooled plates facing upward:[1]

$$Nu_f = 0.27 \, Ra^{1/4} \text{ for } 3 \times 10^5 \leq Ra \leq 3 \times 10^{10} \tag{10-5}$$

The natural convection coefficients for other geometries are given by McAdams[1] and Cess.[2]

From the heat transfer coefficient for natural convection, curve AB in Figure 10-1,

$$q'' = h(T_w - T_{sat}) \tag{10-6}$$

In *nucleate boiling*, two separate processes that are observed are the formation of bubbles (nucleation) and the subsequent growth and flow of bubbles. Nucleation may be either of the homogeneous or heterogeneous type. The metastable states of nucleate boiling are illustrated in Figure 10-3.

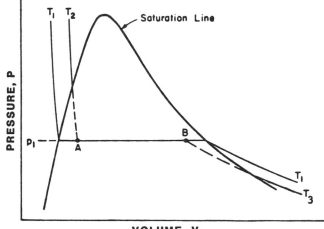

Figure 10-3. Metastable states for liquid and vapor.

Point A denotes the superheated liquid state where the liquid is heated above the saturation temperature T_1 for the constant system pressure P_1 (referred to as $T_2 - T_1$ degrees of liquid superheat). For the corresponding subcooled vapor state B, it is referred to as $T_1 - T_3$ degrees of vapor subcooling. Nucleation theory concerns evaluation of the superheat required to initiate vapor formation. The required superheat is a function of the liquid's temperature gradient, the dissolved and noncondensible gases present in the liquid, surface wetting properties, and the size, shape, and number of cavities present on the heated surface.

Incipient boiling further promotes the bubble growth due to excess vapor pressure. Knowledge of bubble inception, growth, and departure is required to predict the heat transfer performance of a boiling surface. Principal parameters that play a role in bubble formation and dynamics are: fluid state, surface condition, heater and pool geometry, body forces, method of heating, and hysteresis effects. The most widely used correlation for heat transfer in fully developed nucleate boiling is the Rohsenow[4] equation:

$$\frac{C_{p,\ell}(T_w - T_{sat})}{i_{\ell g}} = C_{sf}\left[\frac{q''}{\mu_\ell i_{\ell g}}\left(\frac{g_c\sigma}{g(\rho_\ell - \rho_g)}\right)\right]^n Pr_\ell^m \qquad (10\text{-}7)$$

where C_{sf} is a constant that reflects the condition of a particular liquid and surface combination. Values of C_{sf} are given in Table 10-3. Exponents $n = 0.33$, and $m = 1.0$ for water; $m = 1.7$ for other liquids. Fluid properties in Equation 10-7 should be based on the saturation temperature.

The *critical heat flux* portion of the curve in Figure 10-1 for the pool boiling of water at atmospheric pressure from a flat surface, can only be obtained in its entirety under circumstances where the temperature of the heating surface is carefully controlled. In many cases, however, the surface heat flux is the independently controlled variable. In this case, the boiling curve in the natural convection (AB) and nucleate boiling (B'C) regions remains unaltered. By attempting to increase the value of the surface heat flux above D, the surface temperature jumps from that corresponding to D ($\sim 135°C$) to that corresponding to point D' ($\sim 1150°C$), the next stable operating point in the film boiling region. In practice, this large temperature jump can cause failure of the heating surface (known as "burnout").

The semi-empirical correlation of Zuber[4] can be used for this region:

$$q''_{crit} = 0.18i_{\ell g}\rho_g^{1/2}[\sigma gg_c(\rho_\ell - \rho_g)]^{1/4}[(\rho_\ell + \rho_g)/\rho_\ell]^{1/2} \qquad (10\text{-}8)$$

Table 10-3
Values of C_{sf} for the Rohsenow Correlation
for Nucleate Boiling[3]

Liquid-Surface Combination	C_{sf}
Water–nickel	0.006
Water–platinum	0.013
Water–polished copper	0.013
Water–lapped copper	0.015
Water–scored copper	0.0068
Water–brass	0.006
Water–nickel and stainless steel	0.013
Water–ground and polished stainless steel	0.0080
Water–Teflon pitted stainless steel	0.0058
Water–chemically etched stainless steel	0.0133
Water–mechanically polished stainless steel	0.0132
Carbon tetrachloride–copper	0.013
Benzene–chromium	0.010
n-Pentane–chromium	0.015
n-Pentane–polished copper	0.015
n-Pentane–emery rubbed copper	0.0074
n-Pentane–lapped copper	0.0049
n-Pentane–polished nickel	0.013
Ethyl alcohol–chromium	0.0027
Isopropyl alcohol–copper	0.0025

In the *transition boiling* region, liquid periodically contacts the heating surface resulting in the formation of large amounts of vapor that force the liquid away from the surface and form a vapor film or blanket. The region is normally obtained by controlling the surface temperature to a predetermined value. For water it is observed at atmospheric pressure in a temperature range of about 140°–250°C. Because of the periodic nature of the process, the surface heat flux and temperature undergo large variations of temperature with time, and no adequate model exists.

The lower limit of *stable film boiling* (Leidenfrost point) corresponds to the breakdown of liquid-solid contact and the onset of the continuous insulating film. Analyses are generally based on hydrodynamic stability theory similar to that for evaluating the critical heat flux. For a flat horizontal surface, the minimum heat flux is:

$$q''_{min} = Ci_{\ell g}\rho_g \left[\frac{\sigma g g_c (\rho_\ell - \rho_g)}{(\rho_\ell + \rho_g)^2} \right]^{1/4} \tag{10-9}$$

where $C = 0.09 \sim 0.177$.

Correlations for q''_{min} are unreliable when the liquid contains impurities and the surface has some contamination. Oxidation and increased wettability produce significant increases in q''_{min}.

In *film boiling*, at high temperature differences, a continuous vapor film blankets the heater surface. The major resistance to heat transfer is confined to this vapor film. Vapor is removed from the film in the form of bubbles. The bubble wavelength is that for Taylor instability:

$$\lambda_c = 2\pi \left[\frac{\sigma}{g(\rho_\ell - \rho_g)} \right]^{1/2} \tag{10-10}$$

The relationships for the heat transfer coefficient in laminar or turbulent film boiling in various geometrical situations can be established by analogy with the relationships for filmwise condensation. For laminar film boiling from a horizontal tube:[5]

$$h = 0.62 \left[\frac{g(\rho_\ell - \rho_g)\rho_g k_g^3 i'_{\ell,g}}{d_i \mu_g \Delta T} \right]^{1/4} \tag{10-11}$$

where $i'_{\ell,g}$ is the effective latent heat of vaporization allowing for the effect of superheat:

$$i'_{\ell g} = i_{\ell g} \left[1 + 0.68 \left(\frac{C_{p,g}\Delta T}{i_{\ell,g}} \right) \right] \tag{10-12}$$

Equation 10-11 cannot be applied to very large tubes, horizontal flat surfaces, and very thin wires. For large-diameter tubes and flat, horizontal surfaces, the characteristic length is more correctly λ rather than the tube diameter d_i.[6]

$$h = \left[0.59 + 0.069 \frac{\lambda_c}{d_i} \right] \left[\frac{g(\rho_\ell - \rho_g)\rho_g k_g^3 i'_{\ell,g}}{\lambda_c \mu_g \Delta T} \right]^{1/4} \tag{10-13}$$

A correlation for turbulent film boiling for vertical plates, horizontal tubes, and spheres is:

$$Nu = 0.15 \ (Ra^*)^{1/3}; \ Ra^* > 5 \times 10^7 \tag{10-14}$$

where Ra^* is the modified Rayleigh number; see Rohsenow.[7]

General Notes on Forced Convective Boiling

Forced convective boiling is also referred to as *flow boiling*. A practical problem in the power and process industries is boiling under pressure in a vertically heated tube (constant heat flux boundary condition). Figure 10-4 shows a vertical tube heated uniformly over its length with a low heat flux and fed with subcooled liquid at its base at such a rate that the liquid is totally evaporated over the length of the tube.

Region A is the *single phase convective heat transfer to the liquid*. At some point along the tube, the conditions adjacent to the wall are such that the formation of vapor from nucleation sites occurs. Initially, vapor formation takes place in the presence of a subcooled liquid (region B), known as *subcooled nucleate boiling*. In the subcooled boiling region B, the wall temperature remains nearly constant a few degrees above the saturation temperature, while the mean bulk fluid temperature increases to the saturation temperature. The amount by which the wall temperature exceeds the saturation temperature is the *degree of superheat, ΔT_{sat}*, and the difference between the saturation and local bulk fluid temperature is the *degree of subcooling, ΔT_{sub}*.

The transition between regions B and C, the *subcooled nucleate boiling region* and the *saturated nucleate boiling region,* is the point at which the liquid reaches the saturation temperature ($x = 0$) found on the basis of heat balance calculations. Subcooled liquid can persist in the liquid core even in the *saturated nucleate boiling* region. Vapor generated in the subcooled region is present at the transition between regions B and C ($x = 0$); thus some of the liquid must be subcooled to ensure that the liquid mixed mean (mixing cup) enthalpy equals that of saturated liquid (i_{ℓ}). This occurs due to the radial temperature profile in the liquid, and the subcooled liquid flowing in the center of the channel will only reach the saturation temperature at some distance downstream of the point $x = 0$.

As the thermodynamic mass quality increases through the saturated nucleate boiling region (region D), a point may be reached where a transition in the mechanism of heat transfer takes place. The process of "boiling" is replaced by evaporation. This transition is preceded by a change in the flow pattern from bubbly or slug flow to annular flow (regions E and F). In the latter, the thickness of the thin liquid film on the heating surface is often such that the effective thermal conductivity is sufficient to prevent the liquid in contact with the wall being superheated to a temperature that would allow bubble nucleation. Heat is removed from the wall by forced convection in the film and transferred to the liquid film-vapor core interface where evaporation occurs. When

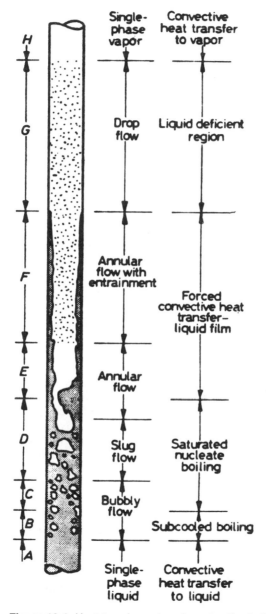

Figure 10-4. Heat transfer regions in convective boiling.[8]

nucleation is suppressed, the system is no longer "boiling." The region beyond the transition is referred to as the *two-phase forced convective region* (regions E and F).

At a critical value of the quality, complete evaporation of the liquid film occurs (known as "dryout"). The area between the dryout point and the transition to *dry saturated vapor* (region H) is called the *liquid deficient region* (region G). Dryout places an effective limit on the amount of evaporation that can be allowed to take place in a tube at a particular value of the heat flux. It is important in the design of evaporators, pipe stills, steam boilers, nuclear reactors, and other units cooled by forced convection boiling.

At relatively low heat flux to the walls, the local heat transfer coefficient can be established by dividing the surface heat flux (constant over the tube length) by the difference between the wall temperature and the bulk fluid temperature. The variation of the heat transfer coefficient with length along the tube for the conditions represented in Figure 10-4 is given in Figure 10-5 (curve (i), solid line). In the *single-phase convective heat transfer region,* the heat transfer coefficient is nearly constant, changing slightly due to the influence of temperature on the liquid physical properties. In the *subcooled nucleate boiling region,* the temperature difference between the wall and the bulk fluid decreases linearly with length up to the point where x = 0. The heat transfer coefficient, therefore, increases linearly with length in this region. In the *saturated nucleate boiling region,* the temperature difference and the heat transfer coefficient are constant. Because of the reducing liquid film in the *two-phase forced-convective region,* heat transfer is characterized by an increasing coefficient with increasing length or mass quality. At the dryout point, the heat transfer coefficient is suddenly reduced from a very high value in the forced convective region to a value near to that expected for heat transfer by forced convection to dry saturated vapor. As the quality increases through the *liquid deficient region,* the vapor velocity increases, and the heat transfer coefficient rises correspondingly. Finally, in the *single-phase vapor region* (x > 1), the heat transfer coefficient reaches a steady-state value corresponding to the convective heat transfer of single-phase vapor flow.

Figure 10-5 is a plot of the heat transfer coefficient versus mass quality with increasing heat flux as a parameter (curves i–vii). Figure 10-5B shows the various regions of two-phase heat transfer in forced convection boiling on a plot of heat flux versus mass quality—"the boiling map." Curve i relates to the conditions for a low heat flux being supplied to the walls of the tube. Curve ii shows the influence of increasing the heat flux. Curve iii shows the influence of a further increase in heat flux. Again, subcooled boiling is initiated earlier and the

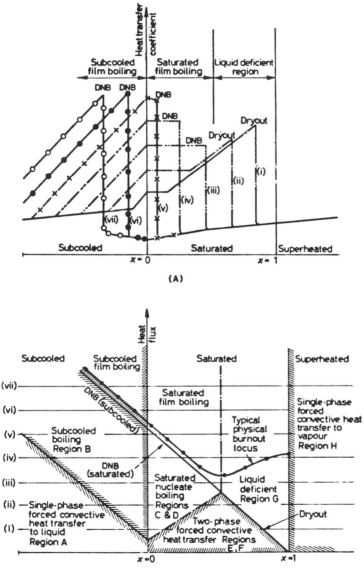

(A)

(B)

Figure 10-5. (A) Variation of heat transfer coefficient as a function of quality with increasing heat flux as a parameter; (B) regions of two-phase forced-convective heat transfer as a function of quality with increasing heat flux as ordinate.[8]

heat transfer coefficient is again higher in the nucleate boiling region. As the mass quality increases before the two-phase forced convective region is initiated and while bubble nucleation is still occurring, an abrupt deterioration in the cooling process occurs. This transition is similar to the critical heat flux phenomenon in saturated pool boiling, called the *departure from nucleate boiling* (DNB).

Film boiling is divided into two regions: *subcooled film boiling* and *saturated film boiling*. Film boiling in forced convective flow is similar to that observed in pool boiling. The heat transfer coefficient is orders of magnitude lower than in the corresponding region before the critical heat flux was exceeded, due mainly to the lower thermal conductivity of the vapor.

Forced Convective Boiling Correlations

The major heat transfer regimes are:

- Single-phase convective heat transfer to liquid
- Subcooled nucleate boiling
- Saturated nucleate boiling
- Two-phase forced convection region
- Single-phase convective heat transfer to vapor

Summaries of heat transfer correlations for each regime follow.

In the *subcooled nucleate boiling* region (i.e., where boiling is first initiated), only a limited number of nucleation sites are operating so that a proportion of the heat will continue to be transferred by normal single-phase processes between patches of bubbles. This transition region is called partial boiling.

The number of nucleation sites increases, and the area for single-phase heat transfer decreases as the surface temperature increases. The entire surface is eventually covered by bubble sites as boiling becomes fully-developed, and the single-phase component reduces to zero. In the *fully-developed boiling region,* velocity and subcooling have little or no effect on the surface temperature. In subcooled boiling, the surface temperature is a function of the surface heat flux and the system pressure for a given fluid. The influence of surface condition should be less for forced convective boiling than for pool boiling because the higher heat fluxes and wall superheats shift the range of active nucleation sites to the smaller sizes, which should be readily available on most surfaces.

Frost et al.[9] report the lower bound for the minimum wall superheat $(\Delta T_{sat})_{ONB}$ as:

$$(\Delta T_{sat})_{ONB} = \frac{4B^{0.5}}{k_\ell}(q'')^{0.5}Pr \tag{10-15}$$

Values for $(4B/k_\ell)^{0.5}$ for various fluids are given in Table 10-4.

Correlations for the fully developed subcooled nucleate boiling regime of forced convective boiling have the same form as those described earlier under the saturated nucleate boiling regime of pool boiling. Thom et al.[10] give the following for *water:*

$$T_w - T_{sat} = 22.65(q'')^{0.5} \exp (-P/87) \tag{10-16}$$

where temperatures are in °C, q'' in MW/m^2 and P in bar.

In the *partial boiling region,* nucleation and the single-phase convection processes occur simultaneously. Rohsenow[11] gives the superposition of a single-phase forced convection component and a subcooled boiling component:

$$q'' = q''_{\ell o} + q''_{SCB} \tag{10-17}$$

where the single-phase convection heat flux is evaluated from:

$$q''_{\ell o} = h_{\ell o}(T_w - T(z)) \tag{10-18}$$

Table 10-4
Values for $(4B/k_\ell)^{0.5}$ Evaluated at a Reduced Pressure of 0.05 according to Collier[8]

Fluid	$\left[\dfrac{4B}{k_\ell}\right]^{0.5}$ (°R/[Btu/hr.ft^2]$^{0.5}$)	Fluid	$\left[\dfrac{4B}{k_\ell}\right]^{0.5}$ (°R/[Btu/hr.ft^2]$^{0.5}$)
Ammonia	0.0138	Oxygen	0.0162
Carbon dioxide	0.0379	Benzene	0.0428
Carbon tetrachloride	0.0436	Water	0.0141
Mercury	0.0127	Freon 12	0.0453
Neon	0.0077	Ethanol	0.0285
n-Pentane	0.0388	Acetone	0.0358
Para-hydrogen	0.0108	Kerosene	0.0405
Nitrogen	0.0166	Helium-4	0.0074
Propane	0.0307	Argon	0.0192

Equation 10-15 can be used to obtain the nucleate boiling heat flux, q''_{SCB}. Bergles et al.[12] outlines a method of constructing the transition between the single-phase convective region and the fully-developed boiling region that is summarized in Figure 10-6. The fully-developed subcooled boiling curve, q''_{SCB}, FED'C" is established from data at high values of $T_w - T_{sat}$ or constructed using empirical equations of the form of Equation 10-16. The single-phase forced-convection curve ABCD' is constructed from Equation 10-18.

The flux, q''_{ONB}, for the onset of nucleation (point C in Figure 10-6) can be calculated from the methods described earlier. Point C establishes the extrapolation of the fully-developed boiling curve at the same surface temperature as point C. An interpolation formula which satisfies the characteristics of the boiling curve between C and E is:

$$q'' = q''_{\ell o}\left\{1 + \left[\frac{q''_{SCB}}{q''_{\ell o}}\left(1 - \frac{q''_{c''}}{q''_{SCB}}\right)\right]^2\right\}^{1/2} \qquad (10\text{-}19)$$

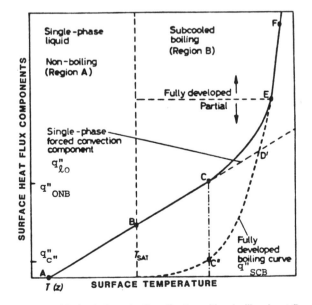

Figure 10-6. Method of evaluating the transition boiling heat flux.[12]

Equation 10-19 provides values of q_{fo}'' and q_{SCB}'' at various T_w for the section CE. This method is applicable for computations of the complete forced-convective subcooled boiling curve.

For *saturated nucleate boiling,* the heat transfer mechanism is essentially identical to that in the subcooled region. The heat transfer coefficient and heater surface temperature variation is smooth and continuous through the thermodynamic boundary ($x = 0$) marking the onset of saturated boiling. The equations used to correlate experimental data in the subcooled region remain valid for this region with $T_\ell(z) = T_{sat}$. As with the subcooled region, the heat transfer process is independent of the "mass quality" $x(z)$ and the mass velocity G. Because the bulk temperature is constant in this region, the heat transfer coefficient is also constant since ΔT_{sat} is fixed for a given heat flux and system pressure.

The *two-phase forced-convective region* is generally associated with the annular flow pattern. Heat is transferred by conduction or convection through the liquid film, and vapor is generated continuously at the liquid film/vapor core interface. Very high heat transfer coefficients are observed. Typical values for water of up to 200 kW/m²°C (35,000 Btu/hr ft²°F) have been reported.

Experimental data on heat transfer rates in the two-phase forced convective region are correlated by an expression of the form:

$$\frac{h_{TP}}{h_{\ell o}} \left(or \ \frac{h_{TP}}{h_\ell} \right) = f\left(\frac{1}{\chi_{tt}}\right) \tag{10-20}$$

where h_{TP} = two-phase heat transfer coefficient
$\quad\quad h_{\ell o}$ = single-phase liquid heat transfer coefficient based on the total flow
$\quad\quad h_\ell$ = heat transfer coefficient based on the liquid component only
$\quad\quad \chi_{tt}$ = Martinelli parameter ($= (\Delta P_\ell/\Delta P_g)^{1/2}$) for turbulent-turbulent flow

A number of relationships of the form of Equation 10-20 have been proposed, and in some cases these have also been extended to cover the saturated nucleate boiling region (see References 7-8).

A correlation that covers both the saturated nucleate boiling region and the two-phase forced convective region follows.[14] It is assumed that both nucleation and the convective mechanisms occur to some degree over the entire range of the correlation and that the contributions made by the two mechanisms are additive:

$$h_{TP} = h_{NCB} + h_c \tag{10-21}$$

where h_{TP} = local heat transfer coefficient
 h_{NCB} = contribution due to nucleate boiling
 h_c = contribution to convection

h_c is the convective contribution given by:[13]

$$h_c = 0.023 \left[\frac{G(1-x)d_i}{\mu_\ell} \right]^{0.8} Pr_\ell^{0.4} \left(\frac{k_\ell}{d_i} \right) F \tag{10-22}$$

"F" is a function of the Martinelli parameter χ_{tt}, and is shown in Figure 10-7.

The equation of Forster et al.[14] can be used as the basis for the evaluation of the "nucleate boiling" component, h_{NCB}. Their pool boiling analysis (Equation 10-15) was modified to account for the thinner boundary layer in forced convective boiling and the lower effective superheat that the growing vapor bubble sees. The modified equation is:

$$h_{NCB} = 0.00122 \left[\frac{k_\ell^{.79} C_{P\ell}^{.45} \rho_\ell^{.49}}{\sigma^{.5} \mu_\ell^{.29} i_{\ell g}^{.24} \rho_g^{.24}} \right] \Delta T_{SAT}^{.24} \Delta P_{SAT}^{.75} S \tag{10-23}$$

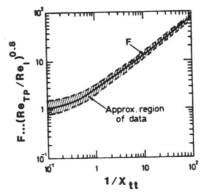

Figure 10-7. Reynolds number factor in Equation 10-22.[13]

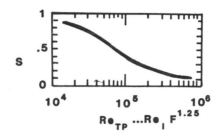

Figure 10-8. Suppression factor S.[13]

where S is the suppression factor defined as the ratio of the mean super-heat seen by the growing bubble to the wall superheat ΔT_{sat}. S is a function of the local two-phase Reynolds number Re_{TP} ($= Re_{\ell}F^{1.25}$) and is given in Figure 10-8.

The correlation in Figure 10-8 fits experimental data with a standard deviation of 11% on the heat transfer coefficient.

To calculate the heat transfer coefficient h_{TP} at a known heat flux (q''), and mass velocity and quality, use the following procedure:

1. Calculate $1/\chi_{tt}$, where

$$\chi_{tt} \simeq \left(\frac{1-x}{x}\right)^{0.9}\left(\frac{\rho_g}{\ell_\ell}\right)^{0.5}\left(\frac{\mu_\ell}{\mu_g}\right)^{0.1} \tag{10-24}$$

2. Evaluate F from Figure 10-7.
3. Calculate h_c from Equation 10-22.
4. Calculate Re_{TP} from $Re_\ell(= G(1-x)d_i)$ and F; $Re_{TP} = F^{1.25}Re_\ell$.
5. Evaluate S from Figure 10-8 using the calculated value of Re_{TP}.
6. Calculate h_{NCB} for a range of ΔT_{sat} values (Equation 10-23).
7. Calculate h_{TP} from Equation 10-21 for the range of ΔT_{sat} values.
8. Plot q'' ($= h_{TP}\Delta T_{sat}$) for the ΔT_{sat} range against h_{TP} and interpolate $(h_{TP})_1$ at q_1''.

Note that Equation 10-22 for the saturated forced-convective boiling regions can be used for all single-component nonmetallic fluids. The correlation is satisfactory for horizontal conduits, provided all the surfaces of the channel remain wetted and stratification is not severe.

References

1. McAdams, W. H., *Heat Transmission,* 3rd edition, McGraw-Hill Book Co., New York (1954).
2. Cess, R. D., "Free-Convection Boundary-Layer Heat Transfer," in *Handbook of Heat Transfer,* J. P. Hartnett and W. M. Rohsenow (Eds.), McGraw-Hill Book Co., New York (1973).
3. Rohsenow, W. M., *Trans. ASME,* 74:969–976 (1952).
4. Zuber, N., *Trans. ASME,* 80:711–720 (1958).
5. Bromley, L. A., *Chem. Engrg. Progress,* 46:221–227 (1950).
6. Breen, B. P., and J. W. Westwater, AIChE Reprint No. 19, 5th National Heat Transfer Conference, Houston, TX (1962).
7. Rohsenow, W. M., in *Handbook of Heat Transfer,* J. P. Hartnett, and W. M. Rohsenow (Eds.), McGraw-Hill Book Co., New York (1973), pp. 13–67.
8. Collier, J. G., *Convective Boiling and Condensation,* McGraw-Hill Book Co., New York (1972).
9. Frost, W. and G. S., Dzakowic, ASME Paper No. 67-HT-61 (1967).
10. Thom, J. R. S., et al., *Proceedings of the Institution of Mechanical Engineers,* 3C180:226 (1965–66).
11. Rohsenow, W. M., *Heat Transfer Symposium at University of Michigan—Summer 1952,* Univ. of Michigan Press, pp. 101–150 (1953).
12. Bergles, A. E., and W. M. Rohsenow, *Journ. of Heat Transfer,* 86C:365–372 (1964).
13. Chen, J. C., *Ind. Engrg. Chem. Process Design and Development,* 5:322 (1966).
14. Forster, H. K., and N. Zuber, *AIChE Journ.,* 1:531–535 (1955).

GLOSSARY

Air-to-fuel ratio—For gaseous fuels, the ratio of volumes in the same units. For liquid and solid fuels, it is the ratio of weights in the same units but can be given in mixed units (e.g., ft^3 air/pound or ft^3 air/gallon).

Air-ramming—Method of forming refractory shapes, furnace hearths, or parts by means of pneumatic hammers, using a plastic ramming mix.

Ash—Noncombustible matter. In residual fuel oils, it is mainly inorganic oxides and chlorides.

ASME—American Society of Mechanical Engineers.

Atomizing air—Portion of air supplied to a burner. It is used to break up the oil stream into small droplets.

Available carbon—Carbon not chemically combined with oxygen and is thus available for reaction or combustion.

Batch Furnace—A furnace that is periodically shut down to remove one load and add a new charge.

Black body—Idealized body that absorbs all radiant energy incident upon it and emits maximum possible radiation at a given temperature.

Black body radiation—Theoretical rate of radiation from a black body at a given temperature.

British thermal unit (Btu)—Quantity of energy needed to heat one pound of water from 59°F to 60°F at a standard barometric pressures, = 0.252 kcal = 0.000293 kWh.

Calcining—Roasting, a process of heating minerals to a temperature below their melting point in which moisture and/or other volatiles, are driven off, thus reducing volume. It can be either an oxidizing or a reducing reaction.

Centipoise—Unit of measure for absolute viscosity = one hundredth of a poise.

Centistoke—Unit of measure of kinematic viscosity = one hundredth of a stoke. Kinematic viscosity in centistokes times the specific gravity equals absolute viscosity in centipoises.

Chiller—A unit that cools a fluid with a refrigerant to a temperature below that obtainable using air or cooling water as the heat sink. Common refrigerants are ammonia, freon, ethylene, and propylene; chilled water or brines are less frequently used.

Coke—Solid product, primarily carbon, resulting from destructive distillation of coal or other carbonaceous matter.

Combustion—Burning or rapid oxidation.

Combustion intensity—The ratio of fuel energy input to the flame volume.

Condenser—A unit in which a process vapor is totally or partially converted to liquid. The heat sink used is ordinarily a utility fluid, such as cooling water. A surface condenser refers specifically to shell-and-tube units used for the condensation of exhaust steam from steam turbines or engines. A "direct contact condenser" refers to a unit in which the vapor is condensed by direct contact heat exchange with droplets of water.

Convection burner—Burner designed to produce a flue gas stream with high velocity, enabling convection heat transfer.

Emissivity—Measure of the ability of a substance to radiate energy equal to the ratio (expressed as a decimal fraction) of the radiating ability of a given material to that of a black body.

Emittance—The ability of a surface to emit or radiate energy, as compared with that of a black body.

Emittance factor—Combined effect of the emittances of two surfaces, their areas, and their relative positions.

Enthalpy—Total heat content, expressed in Btu per pound or kcal/hr, above that at an arbitrary reference condition.

Evaporator—Exchangers designed to concentrate water solutions by vaporizing some of the water.

Firebox—A combustion chamber.

Fireclay brick—A refractory brick manufactured almost entirely from fireclay.

Flash point—Temperature at which enough of a material is vaporized to produce a flash of burning vapor.

Forced-circulation evaporators—With vertical forced-circulation evaporators, liquid enters the tube at the saturation temperature in the overhead vapor-liquid separator, but is at a pressure that exceeds that in the separator due to hydrostatic head and pressure drop caused by flow through the heated tube. Temperature rises as the fluid passes through the tube and eventually reaches a section where ebullition starts. At high entering velocities only a small portion of the feed evaporates.

Fusion point—Temperature at which solid and liquid states of a substance can exist together in equilibrium (also called melting or freezing points).

Gross heating value—Total heat obtained from combustion of a specified amount of fuel and its stoichiometrically correct amount of air, both being at 60°F when combustion starts, and the combustion products being cooled to 60°F before the heat release is measured.

Heat content—Sum total of latent and sensible heat stored in a material minus that contained at an arbitrary set of conditions chosen as the reference. Units are Btu/lb, Btu/gal, or Btu/ft^3 if pressure and temperature are specified.

Heat of combustion—Heat released by combustion of a unit quantity of fuel; units are calories or Btu.

Heat of fusion—Heat derived from a unit weight of liquid freezing to a solid or gained by a solid melting to a liquid, without undergoing a change in temperature.

Heat of vaporization—Heat given off by a unit weight of vapor condensing to a liquid or gained by a liquid evaporating to a vapor, without temperature change.

Heater—Broad classification for heat transfer equipment which adds enthalpy, usually without change of phase, to a stream using a utility (nonprocess stream) as a source of heat. The utility can be steam, a vapor heat transfer fluid such as Dowtherm, or a liquid heat transfer fluid such as Humbletherm or hot oil. Occasionally, a process stream of relatively large heat content is used as a utility; examples are the compressor discharge in a heat pump and the effluent stream of high temperature process reactor processes.

Henry's law—The partial vapor pressure of a' component present in small amounts in a solution is proportional to its mole fraction in the solution; $P = kx$ where k = Henry's law constant for substance.

Induced draft—Gas flow caused by the unit's (e.g., the furnace) exit pressure being less than the unit's pressure. It may be produced by natural or artificial means.

Latent heat of fusion—Heat given off by a unit weight of liquid freezing to a solid or gained by a solid melting to a liquid, without a change in temperature.

Latent heat of vaporization—Heat given off by a unit weight of vapor condensing to a liquid or gained by a liquid evaporating to a vapor, without a change in temperature.

LPG—Liquefied petroleum gas, usually propane and/or butane.

Mullite refractories—Refractory materials consisting mainly of mullite ($3Al_2O_3 \cdot 2SiO_2$) crystals; formed either by conversion of one or more of the sillimanite group of minerals or by synthesis from materials using either melting or sintering processes.

Natural-circulation evaporators—In long tube vertical evaporators, natural circulation is used. Much lower entering velocities are used than in forced-circulation vertical units, and hence, a much larger fraction of liquid evaporates.

Natural convection—Free convection; transfer of heat by currents set up in fluids of different densities resulting from temperature differences.

Overall heat transfer coefficient—Coefficient relating heat transferred from one point to another to the temperature difference between the two points and the cross-sectional area of heat transfer and including the combined effects of several resistances in series as in composite walls with surface and film resistances.

Plasma jet—Ionized gas produced by passing an inert gas through a high-intensity arc, causing temperatures up to tens of thousands of degrees Celsius.

Potential flow—Flow of fluid, heat, or electrical charge, the rate of which is directly proportional to the net driving force and inversely proportional to resistance. For heat transfer

$$q = Q/A = \Delta T/R \text{ or } U\Delta T.$$

Radiant burner—Infrared burner designed to transfer a significant portion of the combustion heat in the form of radiation from surfaces of various refractory shapes or alloy screens in a furnace.

Raoult's law—For a liquid mixture in which the molecules of the respective components are of the same size and mix without molecular association, chemical combination, etc., the partial pressure of any component is equal to its vapor pressure in the pure state multiplied by its mole fraction in the solution

$$p_1 = P_1 x_1.$$

where p_1 = partial vapor pressure of component from mixture
P_1 = vapor pressure of pure component at the mixture temperature

Reboiler—A vaporizer that provides latent heat of vaporization to the bottom (generally) of a fractionation tower. There are two general classes of reboilers, those which send both phases to the tower for separation of vapor from liquid and those which return only vapor. The former operate by either natural circulation (usually called thermosyphon) or forced circulation.

Thermosyphon reboilers are the most common type. Horizontal thermosyphons with vaporization on the shell side are commonly used in the petroleum industry while vertical units with in-tube vaporization are favored in the chemical industry. In a thermosyphon reboiler, sufficient liquid head is provided so that natural circulation of the boiling medium is maintained.

Forced-circulation reboilers require a pump to force the boiling medium through the exchanger. This type of reboiler is infrequently used because of the added cost of pumping the reboiler feed but may be required to overcome hydrostatic head limitations and/or circulation problems.

Reboilers that return only vapor to the tower are called kettle reboilers. The operation of kettle reboilers would be best described as pool boiling.

Regenerator—Cyclic heat interchanger that alternately receives heat from gaseous combustion products and transfers heat to air before combustion.

Rotary dryer (kiln)—An inclined rotating drum usually refractory lined and fired with a burner at the lower end. It is used to dry loose products as they roll through by gravity, sometimes lifted, mixed, and exposed by flights or shelves attached to the inside walls.

Saturated steam—Steam at the boiling point for water at the existing pressure.

Saturation temperature—Boiling point of a liquid for the existing pressure.

Sensible heat—Heat, the addition or removal of which results in a change in temperature, as opposed to latent heat.

Sintering—Heat treatment that causes neighboring particles of material to cohere at a temperature below that of complete melting.

Standard air—Air at standard temperature and pressure; 60°F (15.56°C) and 29.92 in. hg (14.696 lb per in.^{2} or 760 mm hg).

Steam generators (waste heat boilers)—Steam generators are a special type of vaporizer used to produce steam as the vapor product. Generally, the heat source is excess heat beyond that which is required for process; thus the common name of "waste heat boiler" for these units. Like reboilers, steam generators can be the kettle, pump-through, or thermosyphon type.

Theoretical air—Stoichiometric air; chemically correct amount of air required for complete reaction or combustion of a given quantity of material.

Thermal conductance—Also called *conductance*, it is the amount of heat transmitted by a material divided by the difference in temperature of the surfaces of the material. Where heat is transferred by several mechanisms through a structure of mean cross-sectional area A_m, conductance equals the gross rate of heat transfer divided by the temperature drop between its faces:

$$k = \Sigma q/\Delta t = k_m A_m/L$$

Thermal conductivity—Measure of the ability of a material to conduct heat; measured in flow of Btu per hour through a unit cross-section or unit thickness with 1°F of temperature difference across this thickness. For refractory and insulation materials typical units are Btu-in./ft^2-hr-°F. Other acceptable units are Btu-ft/ft^2-°F.

Thermal expansion—An increase in volume or linear dimensions caused by heating the material.

Thermal shock—Denotes a sudden temperature change.

Vaporizer—An exchanger that converts liquid into vapor.

Velocity head—Velocity pressure expressed in feet of column of the flowing fluid.

Wall loss—The heat loss from a piece of equipment to or through its walls.

Watt density—Measure of the concentration (with respect to surface area) of heat input, particularly critical in electrically heated oil heaters. Units are watts per square inch.

Watt-seconds—Unit of work or heat equal to that resulting from the use of electricity at the rate of 1 watt per second. Note 1 W-s = 1 Joule.

UNIT CONVERSION FACTORS

Table 1
Heat Flow Conversion Chart

Note: For two-way conversion of heat flow between British thermal units per hour square foot and watts per square meter, locate in the center column the value you wish to convert. Read equivalent Btu/hr-ft^2 on left or W/m^2 on right. Values are based on the following conversion factors: Btu/hr-ft^2 = 0.316998 × W/m^2; W/m^2 = 3.15459 × Btu/hr-ft^2.

Btu/hr-ft^2	Btu/hr-ft^2 or W/m^2	W/m^2	Btu/hr-ft^2	Btu/hr-ft^2 or W/m^2	W/m^2
			100 to 149		
31.700	100	315.46	39.625	125	394.32
32.017	101	318.61	39.942	126	397.48
32.334	102	321.77	40.259	127	400.63
32.651	103	324.92	40.576	128	403.79
32.968	104	328.08	40.893	129	406.94
33.285	105	331.23	41.210	130	410.10
33.602	106	334.39	41.527	131	413.25
33.919	107	337.54	41.844	132	416.41
34.236	108	340.70	42.161	133	419.56
35.553	109	343.85	42.478	134	422.72
34.870	110	347.00	42.795	135	425.87
35.187	111	350.16	43.112	136	429.02
35.504	112	353.31	43.429	137	432.18
35.821	113	356.47	43.746	138	435.33
36.138	114	359.62	44.063	139	438.49
35.455	115	362.78	44.380	140	441.64
36.772	116	365.93	44.697	141	444.80
37.089	117	369.09	45.014	142	447.95
37.406	118	372.24	45.331	143	451.11
37.723	119	375.40	45.648	144	454.26

(Table 1 continued on next page)

Table 1 Continued

Btu/hr-ft²	Btu/hr-ft² or W/m²	W/m²	Btu/hr-ft²	Btu/hr-ft² or W/m²	W/m²
		100–149 Continued			
38.040	120	378.55	45.965	145	457.42
38.357	121	381.71	46.282	146	460.57
38.674	122	384.86	46.599	147	463.72
38.991	123	388.01	46.916	148	466.88
39.308	124	391.17	47.233	149	470.03
		150 to 199			
47.550	150	473.19	55.475	175	552.05
47.867	151	476.34	55.792	176	555.21
48.184	152	479.50	56.109	177	558.36
48.501	153	482.65	56.426	178	561.52
48.818	154	485.81	56.743	179	564.67
49.135	155	488.96	57.060	180	567.83
49.452	156	492.12	57.377	181	570.98
49.769	157	495.27	57.694	182	574.14
50.086	158	498.43	58.011	183	577.29
50.403	159	501.58	58.328	184	580.44
50.720	160	504.73	58.645	185	583.60
51.037	161	507.89	58.962	186	586.75
51.354	162	511.04	59.279	187	589.91
51.671	163	514.20	59.596	188	593.06
51.988	164	517.35	59.913	189	596.22
52.305	165	520.51	60.230	190	599.37
52.622	166	523.66	60.547	191	602.53
52.939	167	526.82	60.864	192	605.68
53.256	168	529.97	61.181	193	608.84
53.573	169	533.13	61.498	194	611.99
53.890	170	536.28	61.815	195	615.15
54.207	171	539.43	63.132	196	618.30
54.524	172	542.59	62.449	197	621.45
58.841	173	545.74	62.766	198	624.61
55.158	174	548.90	63.083	199	627.76
		200 to 5,500			
63.400	200	630.92	602.3	1,900	5,994
79.250	250	788.65	634.0	2,000	6,309
95.099	300	946.38	665.7	2,100	6,625
110.95	350	1,104	697.4	2,200	6,940

Table 1 Continued

Btu/hr-ft^2	Btu/hr-ft^2 or W/m^2	W/m^2	Btu/hr-ft^2	Btu/hr-ft^2 or W/m^2	W/m^2
		200–5,500 Continued			
126.80	400	1,262	729.1	2,300	7,256
142.65	450	1,420	760.8	2,400	7,571
158.50	500	1,577	792.5	2,500	7,886
174.35	550	1,735	824.2	2,600	8,202
190.20	600	1,893	855.9	2,700	8,517
206.05	650	2,050	887.6	2,800	8,833
221.90	700	2,208	919.3	2,900	9,148
237.75	750	2,366	951.0	3,000	9,464
253.60	800	2,524	982.7	3,100	9,779
269.45	850	2,681	1,014	3,200	10,095
285.30	900	2,839	1,046	3,300	10,410
301.15	950	2,997	1,078	3,400	10,726
317.0	1,000	3,155	1,109	3,500	11,041
348.7	1,100	3,470	1,141	3,600	11,357
380.4	1,200	3,786	1,173	3,700	11,672
412.1	1,300	4,101	1,205	3,800	11,987
443.8	1,400	4,416	1,236	3,900	12,303
475.5	1,500	4,732	1,268	4,000	12,618
507.2	1,600	5,047	1,426	4,500	14,196
538.9	1,700	5,363	1,585	5,000	15,773
570.6	1,800	5,678	1,743	5,500	17,350
		6,000 to 51,000			
1,902	6,000	18,928	8,559	27,000	85,174
2,060	6,500	20,505	8,876	28,000	88,329
2,219	7,000	22,082	9,193	29,000	91,483
2,377	7,500	23,659	9,510	30,000	94,638
2,536	8,000	25,237	9,827	31,000	97,792
2,694	8,500	26,814	10,144	32,000	100,947
2,853	9,000	28,391	10,461	33,000	104,101
3,011	9,500	29,969	10,778	34,000	107,256
3,170	10,000	31,546	11,095	35,000	110,411
3,487	11,000	34,700	11,412	36,000	113,565
3,804	12,000	37,855	11,729	37,000	116,720
4,121	13,000	41,010	12,046	38,000	119,874
4,438	14,000	44,164	12,363	39,000	123,029
4,755	15,000	47,319	12,680	40,000	126,184
5,072	16,000	50,473	12,997	41,000	129,338

(Table 1 continued on next page)

Table 1 Continued

Btu/hr-ft²	Btu/hr-ft² or W/m²	W/m²	Btu/hr-ft²	Btu/hr-ft² or W/m²	W/m²
		6,000–51,000 Continued			
5,389	17,000	53,628	13,314	42,000	132,493
5,706	18,000	56,783	13,631	43,000	135,647
6,023	19,000	59,937	13,948	44,000	138,802
6,340	20,000	63,092	14,265	45,000	141,957
6,657	21,000	66,246	14,582	46,000	145,111
6,974	22,000	69,401	14,899	47,000	148,266
7,291	23,000	72,556	15,216	48,000	151,420
7,608	24,000	75,710	15,533	49,000	154,575
7,925	25,000	78,865	15,850	50,000	157,730
8,242	26,000	82,019	16,167	51,000	160,884
		52,000 to 100,000			
16,484	52,000	164,039	24,409	77,000	242,903
16,801	53,000	167,193	24,726	78,000	246,058
17,118	54,000	170,348	25,043	79,000	249,213
17,435	55,000	173,502	25,536	80,000	252,367
17,752	56,000	176,657	25,677	81,000	255,522
18,069	57,000	179,812	25,994	82,000	258,676
18,386	58,000	182,966	26,311	83,000	261,831
18,703	59,000	186,121	26,628	84,000	264,986
19,020	60,000	189,275	26,945	85,000	268,140
19,337	61,000	192,430	27,262	86,000	271,295
19,654	62,000	195,585	27,579	87,000	274,449
19,971	63,000	198,739	27,896	88,000	277,604
20,288	64,000	201,894	28,213	89,000	280,759
20,605	65,000	205,048	28,530	90,000	283,913
20,922	66,000	208,203	28,847	91,000	287,068
21,239	67,000	211,358	29,164	92,000	290,222
21,556	68,000	214,512	29,481	93,000	293,377
21,873	69,000	217,667	29,798	94,000	296,531
22,190	70,000	220,821	30,115	95,000	299,686
22,507	71,000	223,976	30,273	95,500	301,263
22,824	72,000	227,130	30,432	96,000	302,841
23,141	73,000	230,285	30,749	97,000	305,995
23,458	74,000	233,440	31,066	98,000	309,150
23,775	75,000	236,594	31,383	99,000	312,304
24,092	76,000	239,749	31,700	100,000	315,459

Table 2
Temperature Conversion Chart

Note: The center column of numbers in boldface refers to the temperature in degrees, either Centigrade of Fahrenheit, which it is desired to convert into the other scale. If converting from Fahrenheit to Centigrade degrees, the equivalent temperature will be found in the left column, while if converting from degrees Centigrade to degrees Fahrenheit, the answer will be found in the column on the right.

Centigrade		Fahrenheit	Centigrade		Fahrenheit
−273.17	**−459.7**		−53.9	**−65**	−85.0
−268	**−450**		−51.1	**−60**	−76.0
−262	**−440**		−48.3	**−55**	−67.0
−257	**−430**		−45.6	**−50**	−58.0
−251	**−420**		−42.8	**−45**	−49.0
−246	**−410**		−40.0	**−40**	−40.0
−240	**−400**				
−234	**−390**		−37.2	**−35**	−31.0
			−34.4	**−30**	−22.0
−229	**−380**		−31.7	**−25**	−13.0
−223	**−370**		−28.9	**−20**	−4.0
−218	**−360**		−26.1	**−15**	5.0
−212	**−350**		−23.3	**−10**	14.0
−207	**−340**		−20.6	**−5**	23.0
−201	**−330**		−17.8	**0**	32.0
−196	**−320**				
−190	**−310**		−17.2	**1**	33.8
			−16.7	**2**	35.6
−184	**−300**		−16.1	**3**	37.4
−179	**−290**		−15.6	**4**	39.2
−173	**−280**		−15.0	**5**	41.0
−169	**−273**	−459.4	−14.4	**6**	42.8
−168	**−270**	−454	−13.9	**7**	44.6
−162	**−260**	−436	−13.3	**8**	46.4
−157	**−250**	−418			
−151	**−240**	−400	−12.8	**9**	48.2
			−12.2	**10**	50.0
−146	**−230**	−382	−11.7	**11**	51.8
−140	**−220**	−364	−11.1	**12**	53.6
−134	**−210**	−346	−10.6	**13**	55.4
−129	**−200**	−328	−10.0	**14**	57.2
−123	**−190**	−310	−9.4	**15**	59.0
−118	**−180**	−292	−8.9	**16**	60.8
−112	**−170**	−274			
−107	**−160**	−256	−8.3	**17**	62.6
			−7.8	**18**	64.4
−101	**−150**	−238	−7.2	**19**	66.2
−96	**−140**	−220	−6.7	**20**	68.0
−90	**−130**	−202	−6.1	**21**	69.8
−84	**−120**	−184	−5.6	**22**	71.6
−79	**−110**	−166	−5.0	**23**	73.4
73.3	**−100**	−148.0	−4.4	**24**	75.2
−67.8	**−90**	−130.0	−3.9	**25**	77.0
−62.2	**−80**	−112.0	−3.3	**26**	78.8
−59.4	**−75**	−103.0	−2.8	**27**	80.6
−56.7	**−70**	−94.0	−2.2	**28**	82.4

(Table 2 continued on next page)

Table 2 Continued

Centigrade		Fahrenheit	Centigrade		Fahrenheit
−1.7	29	84.2	23.9	75	167.0
−1.1	30	86.0	24.4	76	168.8
−0.6	31	87.8	25.0	77	170.6
0.0	32	89.6	25.6	78	172.4
			26.1	79	174.2
0.6	33	91.4	26.7	80	176.0
1.1	34	93.2			
1.7	35	95.0	27.2	81	177.8
2.2	36	96.8	27.8	82	179.6
2.8	37	98.6	28.3	83	181.4
3.3	38	100.4	28.9	84	183.2
3.9	39	102.2	29.4	85	185.0
4.4	40	104.0	30.0	86	186.8
			30.6	87	188.6
5.0	41	105.8	31.1	88	190.4
5.6	42	107.6			
6.1	43	109.4	31.7	89	192.2
6.7	44	111.2	32.2	90	194.0
7.2	45	113.0	32.8	91	195.8
7.8	46	114.8	33.3	92	197.6
8.3	47	116.6	33.9	93	199.4
8.9	48	118.4	34.4	94	201.2
			35.0	95	203.0
9.4	49	120.2	35.6	96	204.8
10.0	50	122.0			
10.6	51	123.8	36.1	97	206.6
11.1	52	125.6	36.7	98	208.4
11.7	53	127.4	37.2	99	210.2
12.2	54	129.2	37.8	100	212.0
12.8	55	131.0	40.6	105	221
13.3	56	132.8	43.3	110	230
			46.1	115	239
13.9	57	134.6	48.9	120	248
14.4	58	136.4			
15.0	59	138.2	51.7	125	257
15.6	60	140.0	54.4	130	266
16.1	61	141.8	57.2	135	275
16.7	62	143.6	60.0	140	284
17.2	63	145.4	62.8	145	293
17.8	64	147.2	65.6	150	302
			68.3	155	311
18.3	65	149.0	71.1	160	320
18.9	66	150.8			
19.4	67	152.6	73.9	165	329
20.0	68	154.4	76.7	170	338
20.6	69	156.2	79.4	175	347
21.1	70	158.0	82.2	180	356
21.7	71	159.8	85.0	185	365
22.2	72	161.6	87.8	190	374
			90.6	195	383
22.8	73	163.4	93.3	200	392
23.3	74	165.2	96.1	205	401

Table 2 Continued

Centigrade		Fahrenheit	Centigrade		Fahrenheit
98.9	**210**	410	160	**320**	608
100.0	**212**	414	166	**330**	626
102	**215**	419	171	**340**	644
104	**220**	428	177	**350**	662
107	**225**	437			
110	**230**	446			
113	**235**	455	182	**360**	680
116	**240**	464	188	**370**	698
			193	**380**	716
118	**245**	473	199	**390**	734
121	**250**	482	204	**400**	752
124	**255**	491	210	**410**	770
127	**260**	500	216	**420**	788
129	**265**	509	221	**430**	806
132	**270**	518			
135	**275**	527	227	**440**	824
138	**280**	536	232	**450**	842
141	**285**	545	238	**460**	860
143	**290**	554	243	**470**	878
146	**295**	563	249	**480**	896
149	**300**	572	254	**490**	914
154	**310**	590	260	**500**	932

The formulas below may also be used for converting Centigrade or Fahrenheit degrees into the other scales.

$$\text{Degrees Cent., } °C = \frac{5}{9}(°F + 40) - 40$$

$$= \frac{5}{9}(°F - 32)$$

$$\text{Degrees Kelvin, } °K = °C + 273.2$$

$$\text{Degrees Fahr., } °F = \frac{9}{5}(°C + 40) - 40$$

$$= \frac{9}{5} C + 32$$

$$\text{Degrees Rankine, } °R = °F + 459.7$$

Courtesy Ingersoll-Rand Co.

Table 3
Pressure Conversion Units

To Obtain	Multiply No. of By: Atm	Bars	Millibars	Dynes/cm² (Baryes)	cm of Hg(0°C)	in. of Hg(0°C)	in. of H₂O(4°C)
Atm	1	9.869×10^{-1}	9.869×10^{-4}	9.869×10^{-7}	1.316×10^{-2}	3.342×10^{-2}	2.458×10^{-3}
Bars	1.013	1	10^{-3}	10^{-6}	1.333×10^{-2}	3.385×10^{-2}	2.490×10^{-3}
Millibars	1013	1000	1	10^{-3}	13.33	33.85	2.490
Dynes/cm²	1.013×10^{6}	10^{6}	10^{3}	1	1.333×10^{4}	3.386×10^{4}	2.491×10^{3}
cm of Hg (0°C)	76.00	75.01	7.501×10^{-2}	7.501×10^{-5}	1	2.540	0.1868
in. of Hg (0°C)	29.92	29.53	2.953×10^{-2}	2.953×10^{-5}	0.3937	1	7.355×10^{-2}
in. of H₂O (4°C)	406.8	4.015×10^{2}	0.4015	4.015×10^{-4}	5.354	13.60	1
kg/m²	1.033×10^{4}	10.195×10^{3}	10.195	1.0105×10^{-2}	136.0	345.0	25.40
lb/in.² (psi)	14.696	14.50	1.450×10^{-2}	1.450×10^{-5}	0.1934	0.4912	3.613×10^{-2}
lb/ft²	2116	2.089×10^{3}	2.089	2.089×10^{-3}	27.85	70.73	5.204
ft of H₂O	33.9	33.456	3.3456×10^{-2}	3.3456×10^{-5}	0.4461	1.1329	8.33×10^{-2}
mm of Hg	760	750	0.750	7.50×10^{-4}	10	25.399	1.868
Torr	760	750	0.750	7.504×10^{-4}	10	25.399	1.868
Microns	760×10^{3}	750×10^{3}	0.750×10^{3}	0.750	10^{4}	25.399	1.868×10^{3}
Newtons/meter² (Pascals)	1.013×10^{5}	10^{5}	10^{2}	10^{-1}	1.333×10^{3}	3.386×10^{3}	2.491×10^{2}

Table 3 Continued

kg/m²	lb/in² (psi)	lb/ft²	ft of H₂O	mm of Hg	Torr	Microns	Newtons-meter² (Pascals)
9.678 × 10⁻⁵	6.804 × 10⁻²	4.725 × 10⁻⁴	2.9486 × 10⁻²	1.3157 × 10⁻³	1.3157 × 10⁻³	1.3157 × 10⁻⁶	9.869 × 10⁻⁶
9.8038 × 10⁻⁵	6.8924 × 10⁻²	4.786 × 10⁻⁴	2.9869 × 10⁻²	1.333 × 10⁻³	1.333 × 10⁻³	1.333 × 10⁻⁶	10⁻⁵
9.8032 × 10⁻²	68.924	4.786 × 10⁻¹	29.869	1.333	1.333	1.333 × 10⁻³	10⁻²
98.07	6.895 × 10⁴	478.8	2.9869 × 10⁴	1333	1333	1.333	10
7.356 × 10⁻³	5.171	3.591 × 10⁻²	2.2409	10⁻¹	10⁻¹	10⁻⁴	7.501 × 10⁻⁴
2.896 × 10⁻³	2.036	1.414 × 10⁻²	0.8822	3.937 × 10⁻²	3.937 × 10⁻²	3.937 × 10⁻⁵	2.953 × 10⁻⁴
3.937 × 10⁻²	27.68	0.1922	11.92	0.5353	0.5353	5.3533 × 10⁻⁴	4.015 × 10⁻³
1	703.1	4.882	304.59	13.5979	13.5979	1.35979 × 10⁻²	0.1020
1.422 × 10⁻³	1	6.944 × 10⁻³	0.4333	1.934 × 10⁻²	1.934 × 10⁻²	1.934 × 10⁻⁵	1.450 × 10⁻⁴
0.2048	144.0	1	62.4	2.785	2.785	2.785 × 10⁻³	2.089 × 10⁻²
3.2808 × 10⁻³	2.3076	1.6018 × 10⁻²	1	4.459 × 10⁻²	4.459 × 10⁻²	4.459 × 10⁻⁵	33.46 × 10⁻⁴
7.3553 × 10⁻²	51.710	0.3591	22.409	1	1	10⁻³	7.501 × 10⁻³
7.3553 × 10⁻²	51.710	0.3591	22.409	1	1	10⁻³	7.501 × 10⁻³
73.553	51.710 × 10³	359.1	22.409 × 10³	1000	1000	1	7.501
9.807	6.895 × 10³	47.88	2986.9	133.3	133.3	0.1333	1

Table 4
Conversion Units for Thermal Conductivity (k)

To Convert to Units of	Btu hr-ft-°F	W cm-°C	cal s-cm-°C	kcal hr-m-°C	Btu-in hr-ft²-°F
Btu/hr-ft-°F	1	57.793	241.9	0.6722	0.08333
W/cm-°C	0.01730	1	4.186	0.01171	1.442×10^{-3}
cal/s-cm-°C	4.134×10^{-3}	0.2389	1	2.778×10^{-3}	3.445×10^{-4}
kcal/hr-m-°C	1.488	86.01	360	1	0.1240
Btu-in./hr-ft²-°F	12	693.5	2,903	8.064	1

Multiply These Units by Table Values

Table 5
Conversion Units for Heat Transfer Coefficient (h)

To Convert to Units of	Btu/hr-ft²°F	W/cm²-°C	cal/s-cm²-°C	kcal/hr-m²°C
Btu/hr-ft²-°F	1	1,761	7,376	0.20489
W/cm²-C-°C	5.6785×10^{-4}	1	4.186	1.163×10^{-4}
cal/s-cm²-°C	1.356×10^{-4}	0.2391	1	2.778×10^{-5}
kcal/hr-m²-°C	4.8826	8,600	36,000	1

Multiply these Units by Table Values

Table 6
General Unit Conversions for Heat Flux (q/A)

To Convert to Units of	Btu ft²-hr	W/cm²	kcal/hr-m²	cal/s-cm²
Btu/ft²-hr	1	3,170.75	0.36865	13,277.26
W/cm²	3.154×10^{-4}	1	1.163×10^{-4}	4.1868
kcal/hr-m²	2.7126	8.600	1	2.778×10^{-5}
cal/s-cm²	7.536×10^{-5}	0.2389	36,000	1

Multiply These Units by Table Values

INDEX